# The Springer Series on Human Exceptionality

**Series Editors**

Donald H. Saklofske, Ph.D.
Division of Applied Psychology
University of Calgary, Canada

Moshe Zeidner, Ph.D.
Center for Interdisciplinary Research on Emotions
Department of Human Development and Counseling
Haifa University, Israel

For other titles published in this series, go to
www.springer.com/series/6450

R. Grant Steen

# Human Intelligence and Medical Illness

## Assessing the Flynn Effect

 Springer

R. Grant Steen
Medical Communications Consultants, LLC
103 Van Doren Place
Chapel Hill, NC 27517
USA
G_Steen_MediCC@yahoo.com

ISSN 1572-5642
ISBN 978-1-4419-0091-3          e-ISBN 978-1-4419-0092-0
DOI 10.1007/978-1-4419-0092-0
Springer New York Dordrecht Heidelberg London

Library of Congress Control Number: 2009929164

Printed on acid-free paper

Springer is part of Springer Science+Business Media (www.springer.com)

*To Ralph O. Steen,*
*My father and earliest intellectual guide;*
*To Raymond K. Mulhern,*
*A good friend and valued mentor;*
*To Donald O. Hebb,*
*The finest professor I've ever had.*

# Preface

As critics will note, psychometric tests are deeply flawed. Person-to-person differences in performance on a psychometric test are not informative about many things of great interest. An intelligence quotient (IQ) cannot characterize creativity or wisdom or artistic ability or other forms of specialized knowledge. An IQ test is simply an effort to assess an aptitude for success in the modern world, and individual scores do a mediocre job of predicting individual successes.

In the early days of psychology, tests of intelligence were cobbled together with little thought as to validity; instead, the socially powerful sought to validate their power and the prominent to rationalize their success. In recent years, we have obviated many of the objections to IQ that were so forcefully noted by Stephen Jay Gould in *The Mismeasure of Man*. Nevertheless, IQ tests are still flawed and those flaws are hereby acknowledged in principle.

Yet, in the analysis that follows, individual IQ test scores are not used; rather, average IQ scores are employed. In many cases – though not all – an average IQ is calculated from a truly enormous sample of people. The most common circumstance for such large-scale IQ testing is an effort to systematically sample all men of a certain age, to assess their suitability for service in the military. Yet, it is useful and prudent to retain some degree of skepticism about the ability of IQ tests to measure individual aptitudes.

What follows is an effort to clarify confusion, identify ignorance, and delimit real knowledge. Unfortunately, the confusion and ignorance are often my own, and the real knowledge is usually the work of someone else. I have tried to acknowledge the work of others throughout the book, but some people made contributions so fundamental to my work that those contributions cannot be individually cited. To them, I dedicate this book with gratitude.

Chapel Hill, NC                                                                                        R. Grant Steen

# Contents

# Chapter 1
# Are People Getting Smarter?

We have gone from solitary computers to a buzzing internet in a few years, from horse-drawn carriages to motor-driven cars in a few generations, from standing on the ground to landing on the moon in a century, from tribes to nations in a millennium, from the law of the jungle to the rule of the law in a few millennia. If one thinks of human culture as a kind of collective intelligence, then it is clear that our collective intelligence is growing very rapidly.

The rate of human progress is simply astounding, especially when compared with the rate of cultural change in other creatures widely recognized as intelligent. While we have formed tribes, then kingdoms, then nations, and finally alliances between nations, chimpanzees have continued to live in small family groups. While we have built houses, then towns, then cities, and finally metropolises, dolphins have built nothing tangible at all. While we have learned to speak, then to debate, then to write, and finally to collect writing in libraries, elephants have not progressed past the level of fairly simple vocalizations. While we have made hollowed logs, then violins, and finally symphonies, whales have continued to sing their same simple songs.

The rapidity of change in the human culture argues that the brain itself – the substrate of our intelligence – must also be changing. Yet, there is other and far more direct evidence that the brain is changing; human intelligence appears to be increasing.

## What Is Intelligence?

Intelligence can be hard to define, though most of us can recognize it when we see it. But, what is it that we recognize? The *American Heritage Dictionary* defines intelligence as, "the capacity to acquire and apply knowledge." This definition perhaps places too much weight on acquiring knowledge; computers are far more effective than brains at storing data, yet the storage of data is not intelligence. The ability to apply knowledge is more relevant to what most people would recognize as intelligence, though this definition begs the question: How is knowledge applied? *Stedman's Medical Dictionary* defines intelligence as the "capacity to act purposefully, think rationally, and deal effectively with [the] environment, especially in relation to....meeting challenges."

R.G. Steen, *Human Intelligence and Medical Illness*, The Springer Series
on Human Exceptionality, DOI 10.1007/978-1-4419-0092-0_1,
© Springer Science+Business Media, LLC 2009

This definition seems more substantive, but it also suggests a simpler and equally substantive definition; intelligence is the ability to solve problems.

If human intelligence is simply the ability to solve problems, then it should be possible to measure intelligence objectively. All that would be necessary is to give large numbers of people the same set of problems, then determine who solved those problems more accurately or more rapidly. This is essentially what modern intelligence tests do; they pose a set of relatively simple problems, often in the context of a stringent time limit, and the answers given to these problems are objectively compared to the answers given by a normative sample of people of the same age. Yet, this seemingly simple approach to intelligence testing is fraught with problems.

All intelligence tests assume some degree of prior knowledge; it is not possible to score well in most tests if you cannot read. But, modern intelligence tests strive to measure aptitude rather than achievement – the ability to learn and manipulate new information rather than the ability to recall old information. In other words, intelligence tests strive to gauge an inherent ability to solve new problems, rather than to assess an acquired ability to solve old problems. The logic is, learning the answers to problems in school may make you educated but it does not make you intelligent. The dichotomy between aptitude and achievement is, to some extent, meaningless; no test is completely free of assumed prior knowledge. One cannot exhibit mathematical reasoning unless one has learned the numbers and had some practice with them in the past. Yet intelligence testing strives to be as free as possible from a dependence upon prior knowledge; the era is long past when intelligence might be judged simply by the degree of facility in speaking Latin or Greek.

A difficulty arises when the knowledge base that an intelligence test assumes comprises information that intrinsically favors one test-taker over another, or perhaps favors one culture over another. For example, during World War I, when new immigrants from Italy, Poland, and other parts of Eastern Europe were inducted into the United States Army, the intelligence of these men was evaluated by asking them to recall the nicknames of professional baseball teams [1]. Those who could not speak English were shown drawings and asked to identify the missing element of the picture. If a new immigrant was unable to identify that the net was missing from a drawing of a tennis court, he was assumed to be intellectually inferior. This is clearly unfair to someone who may have grown up in a culture in which tennis is unknown. Yet it is also true that some degree of cultural bias is inescapable in intelligence testing; culture is as ingrained as language and both are learned. Nevertheless, intelligence tests are designed to predict success in the dominant culture, rather than in an alternate culture, so small amounts of bias may not hinder the ability of a test to predict success in that dominant culture.

## The Flynn Effect

Intelligence testing was first undertaken on a large scale by the United States Army during World War I, in an effort to match newly-drafted men with tasks suitable to their skills. Clearly, it would not be appropriate to assign an illiterate man to the officer corps, since an illiterate could neither generate nor respond to written orders. But it would also not be appropriate to put an innumerate man in the artillery corps, since the calculations that enable cannon shells to be fired accurately can be diffi- cult. The Army took up intelligence testing on a large scale to match the man to the mission, to fit the tool to the task.

To this day, some of the best sources of data about human intelligence come from tests given by the military, not only in the United States but elsewhere in the world. Such datasets are more or less perfect for research; very large numbers of men who represent a cross-section of the entire population may be given the same test at the same age and data may be collected and maintained for many years. It was in examining such data that a mysterious trend was first noticed – a trend that has since been called the Flynn effect, after the man who first described the finding in 1984.

The Flynn effect is the tendency for average intelligence test scores to show a progressive increase over time [2]; in other words, people seem to be getting smarter. This effect is so strong that intelligence tests must be rewritten or "renormed" every so often, otherwise people who can perform at the "genius" level would become too common. This is a startling notion to anyone who may have thought that intelligence quotient or "IQ" is immutable. According to scientists at The Psychological Corporation, the company which publishes the Wechsler Adult Intelligence Scale and other widely-used IQ tests:

> All [IQ] tests need revision from time to time and the Wechsler scales are no exception. One of the primary reasons is that scores become inflated over time and norms need to be reestablished. This phenomenon is well documented and has been referred to as the Flynn effect (Flynn, 1984). Therefore, one of the primary reasons for revising the [test] was to develop current norms that would give more precise scores for individuals. [3]

This dispassionate and academic-sounding paragraph downplays what is a truly startling finding. Standardized tests that have in one form or another been given to people for nearly 90 years show that there is a progressive increase in the intelli- gence quotient (IQ) of the average person. This effect is not limited to the United States (Table 1.1), it is not unique to one particular type of test, and it is not some- thing that has been described by one scientist working alone who was promptly contradicted by a host of mainstream scientists. The Flynn effect has been described in a dozen countries – both developed and developing – and in divergent cultures, using a range of different tests [2, 4–18].

How large is the Flynn effect? The answer to this question is contentious, as many different studies have come up with somewhat different answers. But every study of healthy normal people agrees that the average increase in full-scale IQ is on the order of a few IQ points per decade. The studies summarized in Table 1.1 yield two different average values for the Flynn effect, depending upon how the average is calculated. If each study is given identical weight, irrespective of how many people were evaluated in the study, the average increase is 5.2 IQ points per decade. This method of calculating an average assumes that every study is equally likely to be flawed and that enrolling a large number of people does not lead to a more accu- rate estimate of global IQ increase. However, this does not seem to be a reasonable assumption, since the largest study in this compilation was over a thousand-fold

**Table 1.1** Summary of 16 studies in 12 countries, including a total of over 1.4 million people tested with a variety of normed intelligence tests. Estimated probability is the odds that such results could be obtained by random chance alone (0.00001 is 1 chance in 100,000)

| Author | Publication date | Country of study | Sample size | Population description | Average age (years) | Follow-up interval (years) | Change in IQ | Estimated probability | Cognitive test used | IQ points per decade |
|---|---|---|---|---|---|---|---|---|---|---|
| Flynn | 1998 | Israel | 555,339 | Armed forces recruits | 17.5 | 13 | +8.6 | 0.00001 | General intelligence | +6.6 |
| Randhawa | 1980 | Canada | 230,000 | Students | 10–14 | 20 | +7 | 0.00001 | Otis IQ | +3.5 |
| de Leeuw | 1984 | Netherlands | 200,000 | Army recruits | 18 | 30 | +21.1 | 0.00001 | Ravens matrices | +7.0 |
| Rist | 1982 | Norway | 150,000 | Army recruits | 19 | 23 | +10 | 0.00001 | Ravens matrices | +4.3 |
| Teasdale | 2000 | Denmark | 130,621 | Male Army recruits | 18 | 41 | +10 | 0.00001 | General intelligence | +2.4 |
| Bouvier | 1969 | Belgium | 102,400 | Army recruits | 18 | 9 | +5.6 | 0.001 | Ravens matrices | +6.2 |
| Elley | 1969 | New Zealand | 30,000 | Students | 10–13 | 32 | +7.7 | 0.00001 | Otis IQ | +2.4 |
| Clarke | 1978 | Canada | 8,284 | Students | 8–9 | 21 | +8 | 0.001 | Ravens matrices | +3.8 |
| Flynn | 1984 | United States | 7,500 | Whites | 2–70 | 46 | +14 | 0.0001 | Stanford or Wechsler | +3.0 |
| Uttl | 2003 | North America | 7,151 | College students | 20–73 | 31 | ~4 | 0.0001 | Wechsler vocab | +3.7 |
| Vroon | 1984 | Netherlands | 5,694 | Fathers and sons | 18 | 28 | +18 | 0.001 | Ravens matrices | +6.4 |
| Colom | 2003 | Spain | 4,498 | Teenagers | 15–18 | 20 | ~5 | NA | Culture-Fair IQ | +2.5 |
| Lynn | 1987 | England | 1,029 | Students | 9–11 | 50 | +12.4 | 0.001 | Cattell's Nonverbal | +2.5 |
| Daley | 2003 | Kenya | 655 | Students | 7.4 | 14 | +26.3 | 0.0001 | Ravens matrices | +18.8 |
| Fuggle | 1992 | England | 445 | Children | 5 | 21 | 8 | 0.001 | Wechsler | +3.8 |
| Girod | 1976 | France | 400 | Army recruits | 18–22 | 25 | +14.6 | 0.001 | Ravens matrices | +5.8 |
| | | **Total subjects = 1,434,016** | | | | **Average follow-up = 26.5 years** | **Average points/ decade = 5.2** | | **Weighted average points/ decade = 5.4** | |

larger than the smallest study. If these studies are instead weighted according to the number of people tested, which assumes that large numbers of subjects will yield a more accurate estimate of the Flynn effect, then a different average can be calculated. The mean weighted increase in IQ is 5.4 points per decade.

To be conservative and completely above reproach, we will assume that the Flynn effect is actually 5 points per decade or one point every 2 years. If this is true, and if the average IQ is now 100 points, then the "average" person in five generations would be a genius by our criteria. Can it be possible that the human intellect is changing so extraordinarily rapidly?

## How Real Is the Flynn Effect?

Table 1.1 is a compendium of 16 published studies of the Flynn effect. This compendium does not include every study ever published, since a number of additional studies with relatively weak data have been excluded. Nevertheless, IQ gains are documented in 12 countries, including the developing nation of Kenya [16] and 11 developed nations. Subjects include children as young as 2 years and adults as old as 73 years. Army recruits are the dominant demographic in this table, so the most common age for subjects who were evaluated was about 18 years old.

Are these changes in average IQ spurious, an artifact of either small samples or large errors in the measurement? This seems highly unlikely; while the smallest study had only 400 subjects, the largest study included over half a million Army recruits [4]. A total of over 1.4 million people have been evaluated, which is a vast number compared to the number of subjects involved in most scientific studies. Furthermore, the interval between the original test and the follow-up was 26.5 years, on average. Thus, the expected IQ difference over the follow-up period should be around 13 IQ points (using the expected value of 5 points per decade). This difference is large enough to be measured with a high degree of surety, since the reliability of the tests – the ability of the tests to yield the same answer if a single subject is tested twice – is fairly high. In short, there is a great deal of evidence that these IQ changes are both real and meaningful. If IQ tests have any validity whatsoever, these changes cannot be a sampling artifact or an error.

Have human beings really gotten substantially smarter over the last century? This seems too good to be true, given our present understanding of how intelligence arises. Cutting-edge science suggests that increasing IQ can only result from changes in genes or changes in the environment – or both. "Nature versus nurture" is not a satisfying explanation for the human intelligence, since this dogma suggests that both nature and nurture are warring for control of our destiny, that either nature *or* nurture must always be dominant. This is simply wrong. Our intelligence is a product of both heredity *and* environment; it is the interaction between these forces that makes us what we are [1]. While we know that human intelligence is highly heritable, that a child's intelligence is at least half determined by the intelligence of his parents, we also know that the environment – or an interaction between genes and the environment – must be responsible for the other half of a child's intelligence. In other words, heredity and environment must interact to produce high or low intelligence.

## What Could Explain the Flynn Effect?

There are seven hypotheses to explain why the measured human intelligence is increasing, which break down into two main categories of explanation; technical and biological. The three technical hypotheses – which we will review first – basically argue that the Flynn effect is not real. The four biological hypotheses concede that the Flynn effect is real and argue that it is the result of some biological change in humans. We will briefly summarize the three technical and the four biological hypotheses here, developing and testing these hypotheses in greater depth in the rest of the book. We do not claim that these are the only hypotheses that explain rising human IQ, although these are the main hypotheses.

## Hypothesis: IQ Tests Tend to Measure Achievement, Not Aptitude

Can rising IQ simply be a function of better teaching methods, better schools, or perhaps just better test-taking strategies? Modern psychometric tests are the result of years of effort by psychologists to strip away as much as possible of the cultural bias that was obviously present in the earlier tests [1]. Scientists strove to develop tests that characterize the aptitude for future achievement, as well as measure past achievement. Aptitude tests attempt to measure abstract thinking and the ability to reason, aspects of "intelligence" that should have an impact on the ability to achieve in the future, even if past learning opportunities have been limited. On the other hand, achievement tests try to take stock of the knowledge and skills that were attained by the test-taker prior to the test. Obviously, this is a crucial distinction; a gifted child growing up in deprived circumstances may have a great deal of aptitude, but not a great deal of achievement. If the dichotomy between aptitude tests and achievement tests is not as clean as we would like it to be, then rising IQ test results may mean that children are simply learning more or learning earlier.

## Hypothesis: IQ Is Not an Accurate Reflection of Intelligence

Do IQ tests measure something that is perhaps correlated with intelligence, but is not really the same thing as problem-solving ability? The greatest weakness of psychology is the testing tools that have been used to measure mental states, traits, and attributes [1]. Scientists have been developing psychological measurement tools for many years and have been unwittingly incorporating their own biases and preconceptions into these tests. Intelligence was the first mental trait measured in any systematic way, which means that there is a longer tradition of mismeasuring intelligence than of any other human trait [20]. Past failures are so numerous and so egregious that the entire field of psychometrics is open to criticism as being a morass of cultural bias and invalid inference. Therefore, it is possible that we are measuring some factor – call it $g$ – that correlates with problem-solving ability so weakly as to be poorly predictive of life success.

## Hypothesis: IQ Tests Do Not Measure "Average" Intelligence of the Population

Are tests applied uniformly to all students or just to a subset of them? One could easily imagine that intelligence tests are used predominantly for children who are considered intelligent; why bother to confirm that a child who is performing poorly in class has a legitimate reason for failing? Proving that a child has a low IQ would merely discourage or stigmatize the child and perhaps validate an inclination to focus effort away from that child, without serving any useful purpose. Perhaps, teachers have learned to use intelligence tests in a more nuanced way, to prove that good students are smart, without bothering to determine if poor students are stupid.

## Hypothesis: Brain Evolution Is Very Rapid and It Is Happening Right Now

Can brain evolution be so rapid as to leave a trace in the standardized tests? It is clear that the brain is a tool that we are still learning to use, but can it be a tool that is still evolving? A great deal of recent evidence suggests that the brain has evolved more rapidly than any other human organ, with demonstrable changes within the last several millennia. While it is perhaps controversial – at least in some quarters – that the human brain has evolved at all, it is not in the least controversial among biologists [19]. Yet the idea that the human brain evolved so rapidly that we are still learning to use it would be contentious in any quarter.

## Hypothesis: Child Development Is Happening Sooner or Faster than in the Past

Is brain maturation happening faster now than in our parents? A child's performance in an intelligence test is compared to the performance of a normative sample of children tested at the same chronological age. Therefore, if child development is happening faster, current children will be more mature intellectually at the same chronological age than children of the past with whom they are compared. For example, a child who is 12 years and 8 months old now may be intellectually comparable to a child who was 13 years and 4 months old at a point 20 years ago, when normative data were gathered. Yet this would have little practical importance if both children grow up to have the same adult IQ. Nevertheless, if maturation rates are faster now – as they may be – this might make it seem that children are smarter now than they were in the past.

## Hypothesis: The Family Environment Is Improving, Thereby Enabling Intellectual Growth

Is the family environment now more supportive of children than it was in the past? Is it possible that current families are better able to foster the cognitive growth of their children than in earlier generations? Are more families intact and more children safe to explore their intellectual potential? Has the increase in the employment rate of parents over the last decade resulted in more resources being available to children? Or does the decrease in teen pregnancy mean that fewer children are raised by mothers who are overwhelmed by poverty? To what extent can rising IQ be attributed to improvements in the family environment of children?

## Hypothesis: Children Are Healthier and Better able to Demonstrate Intellectual Ability

Are children simply healthier now than in the past? If children are hungry or ill or impaired on some way, they would be less able to concentrate and put forth their best effort on an IQ test. If a substantial fraction of children in the "good old days" actually suffered from hunger or illness or a curable impairment, then those children would likely perform poorly in a test of cognitive ability, while children of today might not be so impaired. Is rising IQ perhaps a sign of increasing general health in the modern world?

Over the ensuing chapters we will explore each of these hypotheses in turn, as well as related hypotheses that may be somewhat less likely than these prime contenders. At issue is whether and why average human intellect is growing in capacity? Given the rapid pace of change in the world and the growing ability of the human species to self-annihilate, there can be few more crucial questions open to scientific inquiry.

# Chapter 2
# Human IQ and Increasing Intelligence

The only real evidence we have that childhood intelligence is increasing is that scores on tests of intelligence are increasing. But what if the tests are wrong? What if tests are a poor measure of intelligence or do not measure it at all? Could it be that the standardized tests so widely used do not measure our ability to learn, but rather measure how much we have already learned?

We have defined intelligence as the ability to solve problems, but virtually any problem that is posed assumes some degree of prior knowledge. For example, most adult IQ tests presuppose that test takers know how to read. Many intelligence tests use analogies ("Finger is to hand as X is to foot") to test logical ability, but analogies indirectly test whether a subject has an adequate vocabulary. Similarly, no one could use mathematical reasoning skills unless they have prior knowledge about numbers, and how to use them. Even a brilliant person who is not a native English speaker might perform rather poorly in a timed test given in English – and many IQ tests are timed.

We have postulated three technical hypotheses that could potentially explain the Flynn effect, all of which contend that intelligence tests somehow mismeasure intelligence. But before we explore these hypotheses in depth, it is important to shed some light on the tests themselves.

## What Is an Intelligence Test?

Typically, an intelligence test is a set of problems, often quite long, with a choice of answers provided. Many questions are used, in part, to derive a more accurate picture of the fine gradations of intelligence, and also to test a person's ability to concentrate for long time intervals. Most intelligence tests have several separate sections and certain sections may have stringent time constraints for task completion. Some tests for young children are given verbally, but most IQ tests for adolescents or adults are given in a written format. Multiple-choice tests are generally preferred because they are easier to score objectively and hence, may be fairer. But it is also true that intelligence testing is a multi-million dollar industry with a keen interest in the bottom line, and multiple choice tests are easier to score.

R.G. Steen, *Human Intelligence and Medical Illness*, The Springer Series
on Human Exceptionality, DOI 10.1007/978-1-4419-0092-0_2,
© Springer Science+Business Media, LLC 2009

Intelligence tests are now so ubiquitous that virtually everyone in the United States has taken one, even if they did not realize that the goal was to measure IQ. Such tests are given to children so frequently in some school districts that many parents no longer pay much attention. The most widely used test for children is the Wechsler Intelligence Scales for Children (WISC), while the most common test for adults is the Wechsler Adult Intelligence Scales (WAIS), and there have been many editions of both the WISC (WISC, WISC-R, WISC-III, WISC-IV) and the WAIS (WAIS, WAIS-R, WAIS-III, WAIS-IV). But there are other intelligence tests as well; a partial list of IQ tests used in recent years would include:

- Wechsler Primary and Preschool Scale of Intelligence (WPPSI)
- Stanford-Binet
- Otis
- McCarthy Scales of Children's Abilities
- Woodcock-Johnson Broad Cognitive Ability
- Slosson Intelligence
- Kaufman Assessment Battery for Children (K-ABC)
- Peabody Picture Vocabulary
- Raven's Standard Progressive Matrices

Typically, some questions are easy enough that most people get them right, while a few questions are hard enough that only a few people can answer them correctly. Clearly, it would be useless to include only questions so easy that everyone got them right or so difficult that everyone got them wrong. If there is not a wide range of difficulty in the questions, then people would not be able to get a wide range of possible IQ scores. The difficulty of individual questions is assessed by determining the proportion of a normative sample that got the answer to the question right. Test questions are analyzed over the years, to make sure that questions are still valid, since it is possible that a question that was once hard might become easier over time, or vice versa, if the knowledge base of the population changes.

Because each new test-taker is compared to a normative sample of people, the composition of the normative sample is extremely important. For the WISC-III test, the normative sample was 2,200 children, ranging in age from 6 years-6 months to 16 years-6 months, who were randomly selected from 79 school districts across the United States [1]. The normative group was carefully balanced so that there were 200 children in each of the 11 age-year strata from 6 to 16, with equal numbers of boys and girls in each stratum. The racial composition of each stratum was designed to include children from the major demographic groups in the United States, in proportions that reflect the population in the 1988 U.S. Census survey. All children in regular classes at school were eligible for inclusion in the normative sample, but children in full-time special education classes were excluded. However, children receiving special education resource room ("pull-out") services were included; overall, 7% of children in the normative sample were learning disabled, speech or language impaired, emotionally disturbed, or physically impaired.

A potential problem with IQ tests is that people who take them repeatedly tend to score better than people who have never taken one before. This is called a practice effect and it can be rather large. In the original WISC-R guidebook, the stability of the test was characterized by selecting a group of 303 children who had already taken the test, then testing them again [2]. Scores in the retest sample correlated closely with the original scores – as measured by the correlation coefficient – meaning that the IQ of individual children can be reliably measured by the WISC-R. Yet there was an average 7 point improvement in full-scale IQ between the first and second test, showing a strong practice effect. Because the practice effect is generally so large, this makes it hard to prove that IQ tests actually measure intelligence, since one would not expect intelligence to change by 7 points over a short time. What the practice effect implies is that test-takers can "learn the test" by taking it repeatedly.

Another problem common to virtually all tests is that a few test-takers score so well or so poorly that their performance cannot be properly described [3]. For example, on the WISC-R test, scoring guidelines are given for each possible answer that a child can provide. But if a child answers every single question wrong, the lowest IQ that can be scored is 48, when the average is 100. This is called a "floor effect," to indicate that 48 is a floor below which one cannot score. Similarly, if a child answers every single question correctly, the highest IQ that can be scored is 158. This means that the test is inappropriate for both profoundly retarded and profoundly gifted children, because the floor and ceiling are too close to the average score.

## IQ Testing of Minorities

Tests of IQ can be administered or scored or used in a racist manner [4]. Without doubt, IQ tests have a small amount of cultural bias, and some tests may have a large amount of such bias. Norms may be inappropriate for some minorities, especially if that minority group was not a part of the normative sample for a particular test. Even if a particular minority was included in the normative sample, the sample as a whole may not be a fair comparator for the individual. This would be especially true if minorities were in some way at a disadvantage in test-taking skills; for example, if a high value is not placed on academic achievement, test-takers may be unwilling to put forth the effort required to do well on an IQ test. It is also worth noting that, especially for a young child, an IQ test is a social occasion; if there is a lack of rapport or, worse, outright prejudice on the part of the test administrator, the examinee is unlikely to perform to his or her capacity. And if IQ test results are ever used to defend or reinforce racial stereotypes, then something is gravely wrong.

There is recent evidence that the expectations or beliefs that are imposed upon people can determine how well they perform on an IQ test [5]. In one key study, Raven's Matrices were used to test black and white college students under three different levels of anxiety or "threat." Under normal conditions of threat, each student was given standard test instructions, which emphasize that the matrices are a measure of the ability to think clearly. Under conditions of "high threat," students were told that the matrices were an IQ test and that their performance was being judged. Under conditions of "low threat," students were told that the matrices were a set of puzzles and that scientists wanted feedback about the puzzles, a story that was told in an effort to remove any evaluative interpretation from the situation. Under conditions of standard and high threat, black students performed more poorly than white students but, under conditions of low threat, black and white students performed equally well. This suggests that black students may incorporate the low expectations that are thrust upon them by prejudice and discrimination. In other words [5], "just the implication that a test is intellectually evaluative is enough to diminish performance among African-American respondents."

Is it possible that the mere threat of evaluation could reduce the performance of some students? Some insight into this question may be provided by a study that randomly assigned similar test-takers to a low-status or a high-status group [6]. People assigned to the low-status group tended to score lower on Raven's Matrices than did people randomly assigned to the high-status group. The mechanism for this effect is not known, but it is possible that anxiety could influence the test performance by – for example – reducing the capacity of working memory. And the idea that one's performance on an IQ test could potentially feed into racial stereotypes could certainly provoke a measure of anxiety. Therefore, anxiety or racial stereotyping can potentially play a role in the disparity of test scores achieved between black and white students [5].

Furthermore, there is evidence that parental expectations can also have a profound effect on how well a child performs on standardized tests [7]. Nearly, 900 black and white children were tested for school achievement, and test results were evaluated in context with parental education and family income for each child. It was found that academic success of the children was often determined by factors that relate most directly to parental beliefs and behaviors. If a parent had low expectations, the child tended to perform poorly, whereas children whose parents inculcated an expectation of success were able to perform well. This study was complicated by the fact that children raised in poor families may receive inadequate schooling, and that children may be poor because their parents also had inadequate schooling. This would make it very hard for parents to help their children to achieve. Nevertheless, there is compelling evidence that the home environment plays a major role in whether a child can succeed in school.

There has also been a great deal of concern that inherent racial bias is a major problem on IQ tests [4]. Yet this commonly held belief seems to lack objective evidence to support it. The fact that some minorities score poorly does not *prove* that there is an inherent bias, since minorities can have many reasons for scoring poorly other than low intelligence; inadequate educational access or a depauperate home environment, or ill health could explain the performance gap shown by minorities.

Efforts to prove content bias on IQ tests have largely been disappointing. In a legal case called *Parents in Action on Special Education (PASE)* v. *Joseph P. Hannon*, a judge singled out seven test questions on the WISC-R as culturally biased. A test group of 180 black children and 180 white children from Chicago city schools were asked to answer these specific questions. Despite the large number of children involved, which should have provided a sensitive measure of bias, it turned out that there was no difference in the proportion of black and white children able to answer the questions correctly. One question that had been singled out as being particularly biased against black children proved easier for black children than for white children to answer. African-American judges were no more able to identify cultural bias than Hispanic judges or judges of European origin. If test questions are ranked in order of difficulty, the ranking is very similar for all races. This means that, although there are hard questions, these questions tend to be hard for everyone. This does not mean that tests are free of bias, but it does suggest that the degree of bias has probably been overestimated.

Standardized IQ tests offer a number of strengths that may well overcome the weaknesses noted above [3]. Tests are usually a valid measure of some form of mental ability, since the results of one test tend to correlate well with the results from other IQ tests. Tests are also usually reliable, as a child will tend to score about the same on separate test components. Procedures used to administer modern IQ tests are generally well-standardized and well-documented, and there has been an effort to make test scoring as fair as possible, while still being sensitive to the nuances of an individual child's response. Finally, some IQ tests have been used so extensively that there is a great deal of research available to help in test interpretation [8, 9].

It is possible that certain types of IQ test do a better job of characterizing the intelligence of minority children. General intelligence is often described as having two components which are correlated with each other rather closely; fluid intelligence and crystallized intelligence [10]. The fact that these types of intelligence are correlated may simply mean that people who beat the rest of us on one cognitive skill often outdo us on other cognitive skills as well [11]. Fluid intelligence – which can be considered as the mental power that underlies knowledge acquisition – is measured by tests of problem-solving that minimize reliance on already-learned skills or prior knowledge. This kind of test should be relatively insensitive to changes in the way that children are taught. Crystallized intelligence is measured by tests that emphasize verbal skill and that rely upon prior knowledge more extensively. A test that only characterized crystallized intelligence would be relatively useless, since it would not predict future ability. Rather, it would be totally dependent upon the quality of schooling that a child had already received. Yet there are no tests that measure fluid intelligence only, without at least a component of crystallized intelligence. Educators and psychologists argue endlessly about the extent to which any specific test actually measures fluid intelligence; what looks like basic knowledge to one person may look like bias to another. But it is clear that the ideal test for a minority child would measure fluid intelligence, rather than crystallized intelligence.

Most psychologists agree that some tests do a better job of measuring fluid intelligence than others. For example, Raven's Standard Progressive Matrices are thought to be "culture-fair" because they assume a minimum of prior knowledge. Raven's matrices were designed to measure a person's ability to discern perceptual relations and to reason by analogy, independent of either language or schooling [12]. Raven's matrices can be administered in less than an hour to people ranging in age from 6 years to adult, and they may seem like a type of game. Test items consist of a series of figures with a missing piece; beneath the figure are a number of pieces that could potentially be used to complete the figure, but only one of the pieces would actually fit. After test completion, the raw score that a test-taker achieves is converted to a percentile using a normative sample matched for age, in much the same way that a score is calculated for other IQ tests. Thus, IQ scores obtained from Raven's Matrices are thought to be equivalent to IQ scores measured by other tests, but with less cultural baggage. However, whether a test that measures fluid intelligence is in reality any different from a test that measures crystallized intelligence is open to debate.

## Hypothesis: IQ Tests Tend to Measure Achievement, Not Aptitude

All multiple choice tests – not just IQ tests – can be classified as tests of either aptitude or achievement. If the goal of a test is to assess the degree of mastery of material that a person should already know, this is a test of achievement. If the goal of a test is instead to assess the ability of someone to master material that they have not had a chance to learn already, this is a test of aptitude. Yet, as we have noted, it is not possible to measure aptitude without assuming some degree of prior knowledge, so all aptitude tests necessarily have a component of achievement inherent to them. A good aptitude test will have a small achievement component, or will at least have an assumed knowledge base that is truly shared among all test-takers.

Modern IQ tests are the result of years of effort by psychologists to strip away as much as possible of the cultural bias that was clearly present in the earliest IQ tests [13]. Scientists have strived to develop tests that characterize aptitude for future achievements, rather than just enumerating past achievements. But what if scientists have failed in their efforts? What if IQ tests still do a poor job of distinguishing between aptitude and achievement? This could mean that rising IQ scores are simply a reflection of better teaching methods or better schools.

If IQ tests truly measure aptitude then they should be able to *predict* long-term achievement or "life success." Yet this is not a fool-proof approach; one can imagine that a person who has suffered enough discrimination that their education is impoverished might also experience career impediments. Nevertheless, there is a clear relationship between intelligence – as measured by IQ tests – and later achievement [14]. For example, IQ scores at a young age predict the grades that a child achieves later in school, as well as subsequent performance on achievement tests, and even specific abilities in reading and mathematical knowledge. In one large study, IQs measured in kindergarten predicted reading achievement scores in the 6th grade in Australia [15]. Another study used a questionnaire to assess student knowledge on a range of subjects that are typically not included in a school curriculum, such as art, engineering, etiquette, fishing, foreign travel, health, law, and so on [16]. Knowledge of these subjects, which are arcane to a child, correlated with general IQ, and the correlation of arcane knowledge with IQ ($r=0.81$, out of a possible maximum 1.00) was much higher than the correlation of knowledge with social class ($r=0.41$). This suggests that a facility with "book learning" is predictive of a facility with "life learning."

People who differ in their IQs tend to also differ in their educational and career attainments. This finding has been dismissed as a self-fulfilling prophecy, but that does not seem adequate to explain the correlation between predicted aptitude and proven attainment. For example, IQ test scores predict the number of years of education that a person is likely to attain, perhaps because a person of higher IQ finds it easier to succeed in school. If brothers of differing IQ are compared, the brother with the higher IQ is likely to obtain more education and to have a more lucrative job than the brother with the lower IQ [14]. In a Canadian study of 250 adolescents born at very low birth-weight, IQ was strongly correlated with scholastic achievement [17]. In a Dutch study of 306 same-sex twins, in whom IQ was measured at age 5, 7, 10, and 12 years, IQ was a reliable predictor of academic achievement at age 12 [18]. In a Chinese study of 104 infants, low IQ at age 5 was correlated with low scholastic achievement at age 16 [19]. In a Chilean study of 96 high- and low-IQ children, IQ was the only variable that explained both scholastic achievement and academic aptitude [20]. Psychologists generally agree that IQ can explain between 25 and 60% of the child-to-child variation in academic achievement, although IQ explains a far smaller proportion of job success [21]. If intelligence is a valid predictor of achievement, then IQ is probably a valid measure of intelligence, since IQ can predict achievement. Nevertheless, it is clear that intelligence did not arise so that children could take IQ tests, so one must maintain a somewhat skeptical attitude about the predictive ability of IQ tests.

It has been proposed that children learn test-taking strategies in school and that the apparent increase in IQ merely reflects the fact that children have learned how to "game" the system [22]. For example, most IQ tests have timed sections, so it is imperative that test-takers be aware of the ticking clock and work within the time limits. It would be a poor strategy indeed for a test-taker to focus on a difficult question and thereby fail to answer easier questions that might appear later in the test. Children today may simply be more comfortable with guessing or with skipping questions that they do not understand, so that they can complete questions that they do know.

Yet even if rising test scores reflect better test-taking skills among children in the United States, this would not explain rising test scores in places where students are unlikely to have been coached in how to take the tests. For example, children in rural Kenya showed a very significant increase in scores on the Raven's matrices between 1984 and 1998 [23]. Raw scores on the Raven's matrices rose by 34% in 14 years, which amounts to an IQ increase of about 26 points, although scores on a test called the Digit Span – which measures memory – did not change. Raven's matrices are, as noted, a fairly pure measure of problem-solving ability ("intelligence"), whereas Digit Span is a very pure measure of memory. Thus, the average child in Kenya is better able to reason now than in the recent past, though their short-term memory has not changed. Most children included in this study were from a single tribal group, so this is a much more homogenous sample that we would expect to see in the United States.

In addition to rising scores on the Raven's matrices, Kenyan children also showed an 11% improvement on a test of verbal comprehension [23]. This test is similar to the Peabody Picture Vocabulary test, which is widely used in the United States. Kenyan children, therefore, showed a strikingly significant improvement on tests that measured IQ by focusing on both fluid intelligence (Raven's matrices) and crystallized intelligence (Verbal Meaning). In short, the Flynn effect was observed in a genetically uniform population of children over a relatively short period of time using different types of cognitive tests that did not change at all over the follow-up interval, and these children are unlikely to have been coached in successful test-taking strategies. This is a very convincing demonstration that rising IQ scores are a reality.

What is crucial about the Kenyan study is that scientists were able to rule out several possible reasons why IQ increased so much in just 14 years. School attendance did not change over the course of the study, and children were tested within 4 months of starting school, hence it is improbable that school attendance could account for the rise in IQ. There was a small increase in pre-school attendance by Kenyan children between 1984 and 1998, but the cognitive tests used in this study were unfamiliar to rural teachers, so it is unlikely that the children were taught how to take the test. This landmark study will be discussed in detail later, but the authors noted that, "it might be profitable to pay … more attention to the role of nutrition, parental education, and children's socialization" as a cause of rising IQ scores [23].

In short, IQ tests – while certainly somewhat flawed – tend to measure aptitude as well as achievement. Thus, the Flynn effect cannot be solely attributed to a rise in achievement, even though this could explain some of the progressive increase in IQ scores.

## Hypothesis: IQ Is Not an Accurate Reflection of Intelligence

Do IQ tests measure something that is perhaps correlated with intelligence, but is not really the same thing as real-world problem-solving ability? This is a compelling consideration, since the most significant weakness of psychology in the past was in the testing tools used to measure mental states, traits, and attributes [24]. Scientists have been developing measurement tools for decades, and for decades have unwittingly incorporated their own biases and preconceptions into the tests. Since intelligence is the first mental trait to be measured in a systematic way, this may simply mean that there is a longer tradition of mismeasuring intelligence than of any other human trait [13]. Past failures are so numerous and so egregious that the whole field of psychometrics has been justifiably criticized as a morass of cultural bias and invalid inference. So it is possible that psychologists are measuring some factor – called $g$ – that correlates with the problem-solving ability, but that only weakly predicts intelligence. In fact, Flynn himself has argued that, "IQ tests do not measure intelligence but rather correlate with a weak causal link to intelligence" [25].

Nevertheless, the weight of evidence suggests that IQ is a reasonably good measure of both aptitude and intelligence. This is hard to prove, but we could try a "thought experiment." Let us provisionally accept the idea that Raven's matrices measure aptitude, whereas most other IQ tests measure something more akin to achievement. Hence, one would predict that children who score well on Raven's should not necessarily score well on other IQ tests. But this prediction is false, as performance on the Raven's is closely correlated with performance on a wide range of other tests of intelligence, including verbal analogies, letter series, arithmetic operations, and even the Tower of Hanoi puzzle [26].

Another "thought experiment" to test whether IQ is an accurate reflection of intelligence also starts with the postulate that most IQ tests measure only achievement. Thus, the Flynn effect could result if children simply learned more at school. Let us further postulate that aptitude – which is measured by Raven's matrices – is not a good measure of intelligence. If this reasoning is true, one would predict that the Flynn effect should be small to non-existent when Raven's matrices are used, because Raven's is not sensitive to achievement. Furthermore, one would predict that people who perform well on Raven's matrices would have no greater success in life than people who do poorly on the test. In fact, both of these predictions are wrong. The Kenyan study showed that Raven's matrices can show a strong Flynn effect [23], and Raven's matrices predict achievement about as well as any other psychometric test [14]. These findings imply that aptitude really is a good measure of intelligence.

Many psychologists argue that fluid intelligence, crystallized intelligence, and achievement are all facets of the same jewel, aspects of a general cognitive ability that has been denoted by $g$. Perhaps the most compelling evidence for the overall unity of intelligence is the finding that all kinds of cognitive demands have a similar effect on the brain [27]. A new method known as functional magnetic resonance imaging (or fMRI) is able to delineate the parts of the brain engaged in a particular cognitive task. The fMRI method can be used to make clear anatomic images of the brain while simultaneously highlighting those parts of the brain that receive a fresh inflow of oxygenated blood. Since inflow of fresh blood occurs when there is an increased demand for oxygen, highlighted brain tissue must be working harder. The fMRI method shows that the frontal part of the brain is engaged during problem-solving, but it is also involved in cognitive tasks as diverse as perception, planning, prioritizing, short-term (working) memory, and episodic (factual recall) memory [27].

Thus, the frontal lobes of the brain, which are the seat of our intelligence, are engaged in a range of different tasks that would not, on the face of it, appear to be part of "intelligence." However, intelligence only happens when past achievement can empower future aptitude, when the separate strengths of a mind can synergize. To clarify by analogy, there is no point in building a sports car with a sleek body, a powerful motor, and a racing suspension, if the transmission is unable to shift easily. It may be that a finely-tuned mind is recognized as such because all the separate components work in harmony, whereas a poorly-functioning mind may have discrete or focal weaknesses that impair the performance of the whole.

In short, all of the evidence marshaled to date suggests that aptitude is an accurate reflection of intelligence. This means that explanations for the Flynn effect based on the idea that IQ does not reflect intelligence probably are not true.

## Hypothesis: IQ Tests Do Not Measure "Average" Intelligence of the Population

Are IQ tests given uniformly to all students or are they being used to identify a subset of students who are gifted? One could well imagine that IQ tests might be used predominantly for children whom teachers already believe to be intelligent. There may be little need to confirm that a child who is performing poorly in class has a legitimate reason for failing. Furthermore, many teachers feel that it is stigmatizing or discouraging to identify a child with low IQ, and parents may not want to know if their child has a low IQ. In short, perhaps teachers have learned to use IQ tests in a nuanced way, to prove that good students are smart, without bothering to determine if poor students are lacking in intellect. This hypothesis boils down to the notion that IQ tests are now used more selectively than in the past, as a way to assess children already identified as being "smart." Yet there is not a shred of evidence to support this hypothesis.

The Israeli Defense Forces used the same two tests to assess all new recruits from 1971 until 1984, without altering these tests in any way [28]. Over this time period, a total of more than half a million men and women were tested (Table 1.1) and measured IQ rose at a rate of 6.6 points per decade. Because service in the armed forces is a prerequisite for all Israelis, both men and women, virtually every Israeli male or female was tested when they reached the age eligible for service. The only people who were not tested were institutionalized, terminally ill, crippled, or eliminated by medical screening, and the criteria used for medical screening did not change over the course of the study. In other words, this is not a sample of people selected because their IQ is high; this is a sample which represents virtually every person in a population, including people who were moderately disabled or had severe psychological problems. Nevertheless, in a huge sample that included virtually everyone in the country, the average IQ of the population increased by nearly 9 points within 13 years.

The Danish draft board has assessed all men eligible for service in the Danish Army, using tests that were unaltered from 1957 until 1997 [29]. Only about 10% of men were excluded from cognitive testing for medical reasons, so this sample also represents a cross-section of the entire population, rather than a non-random sample of the "best and brightest." During the 41 year follow-up of this study, a total of over 130,000 men were tested, and the average IQ rose by 10 points, which is relatively small compared to other countries (Table 1.1). Nevertheless, IQ rose in a way that could not be accounted for by selective administration of the test.

Furthermore, some very interesting trends emerged when a closer look was taken at the data. For one thing, the rate of change in IQ seems to be slowing down. The last 10 years of the study showed rather modest gains in IQ compared to the first 10 years of the study. Across the four decades of the study, gains in IQ worked out to be roughly equivalent to 4 points, 3 points, 2 points, and 1 point per decade. In other words, the rate of change in IQ from 1957 until 1967 was four times as fast as from 1987 until 1997.

Another fascinating aspect of the Danish Army data is that changes in IQ were not uniform across the sample [29]. The average gain in IQ for men at the 90th percentile of ability was quite small, working out to be a 4% increase in raw score. But the average gain in IQ for men at the 10th percentile of ability was enormous, working out to be a 71% increase in raw score. While the smartest men were not getting very much smarter, the dullest men were getting dramatically brighter. In 1957, 10th percentile men were performing at just 33% of the level attained by 90th percentile men. In 1997, 10th percentile men were performing at 54% of the level attained by 90th percentile men. Thus, the gap between the brightest and the dullest men had closed considerably. And because the dullest men improved enormously, the average IQ of the group increased by about 21%. In other words, average IQ was far more influenced by changes at the low end of the spectrum than by changes at the high end of the spectrum.

Apparently, no other study has been analyzed in exactly this way, but it seems likely that this pattern would be replicated in other studies in which the average IQ has increased. In short, the Flynn effect may be a function of increasing IQ among those people most impaired by low IQ, rather than among those who are most gifted. This tentative conclusion is supported by careful inspection of several plots of average IQ of Danish men born between 1939 and 1958 [30]. For the cohort of men born between 1939 and 1943, the plot of IQ looks like a standard "bell curve," with an average value at the center and a symmetrical distribution of values around that average. For the cohort of men born between 1954 and 1958, the distribution is skewed, such that there are fewer men in the lower tail of the IQ distribution and more men in the upper tail of the distribution. It is as if the whole bell curve of IQ had been shoved toward the higher IQ. This is really a profound message of hope; increases in IQ seem to occur in exactly that segment of the population that needs a change most desperately.

It is worth emphasizing that, whenever the Flynn effect is discussed, whenever people note that IQ is rising at a rate of 5 points per decade, what is changing is the population IQ. No one imagines that – except in rare cases – the actual IQ of any one individual is changing very much, and certainly individual changes in IQ could not be sustained over decades. Instead, the Flynn effect refers to the aggregate IQ of a large group of people. This is perhaps less exciting than if individual IQ were to change but it is, statistically and scientifically, a far more robust thing. As is true of any measurement, there is a certain amount of error or "squish" in the measurement of individual IQ. If the IQ of an individual person was to increase by a point in 2 years, this could easily be dismissed as a random variation or a practice effect, or the result of "learning the test." But if the IQ of half a million people increased by about 9 points in 13 years [28], as happened in Israel (Table 1.1), this cannot be dismissed as an artifact. Rather, this is a real-world change in the IQ of the average person taking the test in a given year. In Israel, IQ was measured whenever people were old enough for military service, so the IQ measured in any year is the collective IQ of everyone who was then about 18 years old.

Evidence indicates that the technical hypotheses meant to explain the Flynn effect are probably not adequate to do so. Instead, we are left with a sense that only a biological rationale can explain why IQ has been rising for several decades. This of course makes the Flynn effect a far more interesting and challenging problem.

# Chapter 3
# Evolution and Increasing Intelligence

We have seen that various technical or non-biological explanations for the Flynn effect fall short of explaining what appears to be a progressive worldwide rise in IQ. These explanations – that IQ tests really measure achievement rather than aptitude; that IQ is not an accurate measure of intelligence; and that IQ tests do not actually measure the "average" intelligence – are either demonstrably false or inadequate to explain an enormous rise in IQ. This means that we are left with the far more interesting possibility that there is a biological explanation for the Flynn effect.

One possible hypothesis – as farfetched as it may seem – is that human brain evolution is very rapid and it is happening right now. This hypothesis argues that the brain is evolving so rapidly that evolution is literally leaving traces in the form of rising IQ scores. There is clear and abundant evidence that human evolution has happened, especially in the brain, but what evidence is there that evolution is currently going on? In the course of evaluating the main hypothesis of human brain evolution, we will evaluate several subsidiary hypotheses, each of which relate to the human evolution.

## What is Evolution?

Evolution is simply a generational process of change. Gradual alteration of the phenotype – which includes physical appearance, individual behavior, and brain function – occurs as a result of changes in the genotype – those genes that encode the phenotype. Such changes are the result of a fairly uncomplicated but profoundly important process. The following simple principles are necessary and entirely sufficient to explain a gradual change in phenotype that any biologist would recognize as evolution:

1) **Variation exists**. This observation is so simple that it seems incontrovertible. Any large gathering of people will include people who are slim and fat, old and young, tall and short, weak and strong, healthy and ill. It could perhaps be argued that most variations are meaningless in an evolutionary sense, especially in a human gathering, and this may well be true. Yet variation exists in all organisms of all species in all places.

R.G. Steen, *Human Intelligence and Medical Illness*, The Springer Series
on Human Exceptionality, DOI 10.1007/978-1-4419-0092-0_3,
© Springer Science+Business Media, LLC 2009

**2) Some variants are more successful than others**. Imagine a herd of antelope, in which some are slim and some are fat, some are old and some are young, some are weak and some are strong. Clearly, if a lion were stalking that herd, then the fat or the old or the weak would be more likely to die. Predation is not random; lions risk injury every time they hunt, so they always seek the weakest prey. Even then, lions are not always successful, since they are sometimes unable to kill any prey. Nevertheless, over time, there is a stronger selection pressure working against the weak than against the strong.

**3) Variation is heritable**. Everything we know about our own families convinces us that certain traits are likely to run in families: tall parents tend to have tall children, just as near-sighted parents tend to have near-sighted children. Everything we know about genetics concurs that certain traits are passed down to offspring, often with a high degree of fidelity. This simple truth can have terrible consequences, as certain families are ravaged by hereditary illness.

**4) Successful variants tend to become more abundant over time**. Because certain individuals are more likely to survive long enough to reproduce, and because these individuals are able to pass specific traits on to their offspring, these traits will tend to be well-represented in ensuing generations. In contrast, other individuals may have traits that are more likely to lead to premature death, so these traits are less likely to be passed down. Over time, the successful traits will tend to increase in the population, whereas unsuccessful traits will gradually decrease [1].

Evolutionary change emerges from a combination of processes that would seem to be polar opposites: a random process of change and a non-random process of culling the result. Mutation generates few changes that are ultimately successful, just as a blind watchmaker could rarely alter a watch to make it keep better time. Just as we would not accept the alterations of a blind watch-maker without verifying that the watch still works, random changes to an antelope are subject to the stringent selection pressure of a hungry lion. If mutation makes a change to the genome of an organism, that organism must survive long enough to reproduce, in order to pass genetic changes on to later generations. Otherwise, there can be no evolution.

The process of natural selection culls unsuccessful mutations from a population in the most implacable way possible [1]. The lame, the halt, the weak, the maladapted – all are slaughtered with a fierce egalitarianism. Evolution is not a random process at all. Yet it is generally a slow process; each evolutionary "experiment" is evaluated by natural selection over the lifetime of that organism. This evaluation process can be brutally quick – if the mutation makes some disastrous change – but more often it is rather slow. While the process is ultimately inexorable, it is not as effective as one might imagine. For example, heart disease can kill people at a young age, but usually not so young that those with a weakened heart are unable to have children of their own. Thus, there is effectively no selective pressure against heart disease in humans, unless the birth or survival of children is somehow impaired.

Evolution is usually a change in phenotype that occurs over geologic time. Yet evolution can potentially also happen rapidly under certain limited circumstances. We will evaluate these few circumstances, to determine if they are relevant to the progressive increase in human IQ.

## What If There Was Stringent Selection Against the Dull-Witted?

Imagine that a herd of antelope has been under steady selection pressure from the local lions when a new mutation abruptly arises that enables certain antelopes to run somewhat faster or be more elusive. Lions would be just a little less likely to catch these altered antelopes, such that the lion's predation would be concentrated only on those antelopes that lacked the mutation. At first, this greater selection pressure on the slow antelopes would hardly make a difference; the altered animals would be rare and hence, predation would be evenly spread over a large number of "normal" antelopes. But, over time, if the normal antelopes were somewhat less successful in evading the lions, then the altered antelopes would become progressively more common. At some point, the new mutation might become prevalent in the antelope population, which would mean that those antelopes lacking the mutation would become the focus of predation by lions. Eventually, as slower antelopes became a rarity, selection pressure against them would steadily mount, as lions directed all their attention to those few remaining antelopes that could be easily caught. At this point, there would be a stringent selection pressure against being slow-footed.

Similarly, if a human genetic alteration led to some people being smarter, these few might have some selective advantage, yet there would not at first be stringent selection against the dull-witted. However, if smart people underwent a progressive increase in the population, then there might eventually be stringent selection pressure against those people who remained no smarter than their ancestors. In fact, a point could be reached – called the "tipping point" – when the selective pressure against the dull-witted would become fairly strong. For example, if dull-witted people were unable to find sexual partners and to have children, a tipping point could be reached when conditions would change rapidly and the selective pressure against obtuseness might become strong.

Yet this evolutionary fairy tale does not seem to be happening to humans. First, there is no demonstrable selection pressure against low intellect, at least in an evolutionary sense; people of low IQ may not hold the most lucrative jobs, but they are still generally able to feed themselves and to raise a family. Secondly, this mode of evolutionary change is expected to occur slowly, as smart people gradually come to represent a greater proportion of the population; it seems likely that only at the tipping point would change in the average intelligence occur as rapidly as a few points per generation. Yet we cannot be at such a tipping point, because there is no obvious selection pressure against those who are obtuse.

Has there *ever* been stringent evolutionary selection pressure against the dull-witted? There is intriguing evidence that Neanderthals did not show any increase in average intelligence, even when they were subjected to very stringent selection pressure. In fact, Neanderthals may have been driven to extinction by our own brainier ancestors. This evidence should be regarded with caution; clear evidence of intelligence is hard to find in the fossil record, because cranial volume is a surprisingly poor surrogate measure for IQ and there is little else that can fossilize. Whether Neanderthals actually went extinct because they were unintelligent is not known, but their culture was simpler and more primitive than those of our own lineage alive at that time [2]. Neanderthal archaeological sites typically lack art or jewelry, which was common among our ancestors, and there is scant evidence of any burial ritual. Compared to human ancestors, Neanderthals used a smaller range of stone tools, they rarely crafted tools from bone, antler, tusk, or shell, and they lacked any projectile weapons. It is fair to say that Neanderthals were, if not dull-witted, at least less like modern people than were our ancestors, often called Cro-Magnons.

The incompleteness of the fossil record makes any conclusion tentative, yet Neanderthals do not seem to provide an example of increasing intelligence, even in response to stringent natural selection. Neanderthals emerged in a distinctive form in Europe more than 130,000 years ago and they continued to exist there until roughly 30,000 years ago [2]. Research that sequenced DNA extracted from a fossil Neanderthal bone roughly 38,000 years old suggests that the Neanderthals and Cro-Magnons diverged evolutionarily about 500,000 years ago [3]. Neanderthals were dissimilar from Cro-Magnons in that they had a more protuberant lower face, more massive jaws, more robust bones, a slightly larger braincase, shorter limbs, and a far more massive trunk. The weight of evidence thus suggests that the last shared ancestor between Neanderthals and Cro-Magnons was alive at least half a million years ago. While Neanderthals occupied Europe and western Asia, our more modern-looking ancestors occupied Africa and western Asia. But modern humans invaded the eastern part of the Neanderthal's range about 45,000 years ago and gradually swept west through Europe, pushing aside Neanderthals in a wave of immigration that was completed within about 15,000 years. There are no archaeological sites anywhere in the world where Neanderthals and Cro-Magnons were clearly contemporaneous, so Neanderthals may have disappeared abruptly.

Both Neanderthals and Cro-Magnons shared certain behaviors, including an ability to flake stone into tools, an interest in caring for their aged and burying their dead, full control over fire, and a dependence upon large-animal hunting to supply meat for their diet [2]. Both hominids exploited the same food sources and – if they actually coexisted – must have been in competition with one another, as well as with other carnivores such as hyenas [4]. It is not hard to imagine that the technologically inferior Neanderthals, who lacked projectile weapons, must have suffered in competition with Cro-Magnons. There is evidence that Neanderthals survived in a refugium near Gibraltar until as recently as 23,000–33,000 years ago [5]. Gibraltar was then an area of coastal wetland and woodland, and may have been a rich source of birds, reptiles, and shellfish. Our ancestors were perhaps contemporaneous with

Neanderthals near Gibraltar; an archaeological site with human bones from the same era is 100 km away at Bajondillo, Spain. Yet genetic evidence suggests that Neanderthals contributed to no more than 5% of human genes, thus, the wholesale intermixing of genes that might be expected if the two lines merged did not exist [6]. Mitochondrial DNA, extracted from the first Neanderthal bones ever found, has a sequence far outside the range of variation seen in modern humans, suggesting that Neanderthals were replaced rather than assimilated by our ancestors [7]. Fossil evidence concurs that Neanderthals and Cro-Magnons remained largely distinct from each other; teeth, which are the most abundantly preserved type of fossil, are distinctly different in Neanderthals and modern humans [8].

Nevertheless, some scientists have proposed that Neanderthals and modern humans coexisted in a mosaic of small "tribes" that had minimal contact with one another [2]. This scenario seems unlikely, and the latest evidence suggests that the period of overlap may have been even shorter than inferred before [9]. Apparently, errors have been made in the radiocarbon dating of many fossils; samples are easily contaminated by relatively young organic acids in ground water that can percolate into a fossil from the surrounding soil. If radiocarbon ages of Neanderthal and Cro-Magnon bones are recalculated, using corrections for this and other common sources of error, a striking finding emerges. Complete replacement of Neanderthals by Cro-Magnons may have taken as little as 5,000 years, which is more than twice as fast as previously thought. In fact, within specific regions, the period of overlap was perhaps as short as 1,000 years, which suggests that Neanderthals were literally shoved out of existence by Cro-Magnons.

However, even when stringent selection against Neanderthals should strongly favor the ability to develop and use new weapons for hunting, there is little evidence that the culture of Neanderthal changed [2]. In fact, Cro-Magnon culture changed more over time, implying that Neanderthals may have had a limited ability to respond to their changing environment. What this may mean is that Neanderthals were so obtuse that they were driven to extinction by Cro-Magnons. However, when the rate of evolutionary change should have been maximized by stringent selection pressure against the dim-witted, Neanderthal intelligence apparently did not change much at all.

## What If Very Intelligent People Tended to Have More Children?

It should be clear that if very intelligent people had more children, then the brightest people would be better represented in later generations. Higher average intelligence test scores would be expected for the population as a whole, and this upward drift in intelligence would eventually be recognizable as an evolutionary change in the population. The only problem with this idea is that it is simply not happening.

There is an enduring prejudice that unintelligent people have larger families. Whether or not this prejudice contains any hint of truth is irrelevant, because it is exactly the opposite of what is needed to explain the Flynn effect. The fear that people of low intelligence will swamp human progress by producing children at a faster rate than others is completely groundless, since the trend is actually toward increasing human intelligence. This cannot be emphasized too much; fear that the unfit will somehow out-compete the fit – unless the unfit are held in check – is the source of much evil in the modern world. The rise to power of Nazis in Germany was propelled by a belief that a pure Aryan stock was being corrupted by gene mixing with lesser peoples. Yet the Flynn effect was probably happening in Germany well before the Nazis rose to power. Similarly, the fear that immigrants or ethnic minorities are sullying the American fabric cannot be true, because the average IQ is increasing by a point every 2 years in the United States. This observation should be the death of racism, yet racism remains immune to logic.

Even if it were true that very intelligent parents tend to have larger families, this would likely produce little or no change in average intelligence of the population. This is because there is evidence that, all other things being equal, birth into a large family leads to lower intelligence. In what may be a definitive experiment [10], scientists evaluated the intelligence of all children born in Norway between 1967 and 1998. Records were sorted so as to include only males who later took an IQ test when they were inducted into the Norwegian Army, but still 243,939 men were involved. First-born sons were compared to later-born sons, and it was found that first-borns scored about 3 points better on an IQ test than did second-born sons. Compared to third-born sons, first-borns scored more than 4 points higher on an IQ test. Yet this was not the end of the analysis; because data were available from so many men, it was possible to address why first-born sons are smarter. There are two possibilities: one theory is that first-born sons are smarter because they spend more quality time with their parents and hence enjoy a richer social environment. Another theory is that first-born sons are smarter because of some factor related to gestation; perhaps older mothers are less healthy or less able to support growth of a second fetus. To test these competing hypotheses, scientists compared first-born sons to second-born sons who had lost an older sibling in childhood. Such second-born sons would experience a social environment much like what a first-born son would experience, even though they were born to older mothers. Scientists found that, in this situation, there was no significant difference between first- and second-born sons. This strongly implies that any deficits in the intelligence of later-born sons are due to deficits in the social environment, and not due to deficits in the health of the mother. This is a convincing demonstration of what scientists have long suspected; family interactions have an enormous impact on IQ. Yet this study is the first to show clearly that large families can produce a form of social impoverishment that reduces the IQ of later-born children.

In another study, first-born children had an average IQ of 109 points, whereas fifth-born children had an average IQ of 100 points [11]. There was a very significant trend for IQ to decrease as birth order increased, perhaps because parental resources are spread thinner in large families. The trend remained significant even after correction for maternal age, maternal health at delivery, and social class of the parents. Conclusions from this study are probably robust because a huge number of people were evaluated; this study enrolled 10,424 children. In fact, this study enrolled more than 90% of the children born in Aberdeen, Scotland, all of whom received an IQ test at age 7, 9, and 11 years. It should be noted that all of the known predictors of IQ together could explain only about 16% of the variation in a child's intelligence, implying that much about intelligence remains unknown. Nevertheless, the Aberdeen study is an unusually thorough assessment of a large sample of children, with data evaluated by modern methods [11]. These and other findings suggest that, if intelligent parents insist on having large families, it is quite likely that their later-born children would not be as intelligent as they might have been, if born to a smaller family.

The Aberdeen study confirms a great deal of prior work showing that very large families tend to produce children of lower intelligence. For example, an earlier study of 36,000 college applicants in Columbia, South America, found that moderate-sized families produce smarter children than large families [12]. Many studies confirm that smaller families tend to produce children with greater intelligence [13]. The effect is so strong that it has led to the "resource dilution" hypothesis; the idea that, as the number of children in a family increases, parental resources are spread thinner, so that the resources available to any individual child must decrease [14]. Recently, scientists have questioned the validity of prior studies, and some researchers have concluded that birth order has no impact at all on a child's intelligence [15]. This is still a minority view, but most scientists would agree that the impact of birth order on IQ is not crucial [16]. In any case, the relationship between family size and intelligence cannot explain the Flynn effect. For rising general intelligence to be explained by family size, large families would have to be both smarter and increasingly more common, and neither trend is found.

## What If Selection for Intelligence Was Indirect?

Intelligence might undergo an evolutionary increase in the human population if it was closely correlated with some other trait that is itself under stringent selection pressure. Suppose, for example, that human intelligence is in some way genetically linked with resistance to influenza. This would mean that through some unknown mechanism, people with a natural resistance to the flu would tend to be smarter. There was a worldwide influenza pandemic in 1918 that killed more than 20 million people [17] and, no doubt, some people were more vulnerable than others to this illness. Perhaps the 1918 pandemic and later waves of flu selectively killed people of somewhat lower intelligence. If this were true, it could mean that any time the flu tore through a country, the average intelligence of the people might increase somewhat. From an evolutionary perspective this might make some degree of sense, because intelligence could increase without ever being directly the result of natural selection.

Yet this proposed mechanism is wildly implausible, for several reasons. First, if selection for influenza resistance is to result in an increase in human intelligence, there must be a close physical linkage between the gene(s) for flu resistance and the gene(s) for human intelligence. Otherwise, there could be people with flu resistance who are not intelligent, or people who are intelligent but not resistant to flu. For the proposed mechanism to be able to produce a rapid increase in intelligence, it must be true that high intelligence and flu resistance co-occur in the same person frequently. But this cannot be true because an unknown but large number of genes must interact to produce a person of high intelligence. The various "intelligence" genes probably reside on different chromosomes in the human genome, so they cannot be physically linked to the gene(s) that might make a person resistant to influenza. The same logic applies to any other trait that could be indirectly selected; intelligence is due to many genes scattered all over the genome, and it is not possible to imagine a scenario whereby selection for an unrelated trait could produce an increase in IQ. Just as there is no evidence that influenza spared the intelligent, there is likewise no known linkage between IQ and any trait that might be prone to natural selection.

## Hypothesis: Brain Evolution Is Very Rapid and It Is Happening Right Now

The main reason that evolution cannot be invoked to explain the Flynn effect is that evolution is only possible over relatively long periods of time. A change in human structure or function that we would recognize as evolution would take millennia, not decades. There simply has not been enough time passed to attribute the increase in human intelligence to an evolutionary force. Evolution of human intelligence is happening, yet it is happening over a geologic time period, rather than over an historic time period.

Nevertheless, it is possible that evolution can sometimes happen quickly. About 10 years ago, it was recognized that certain people are less likely to develop acquired immune deficiency syndrome (AIDS) even after repeated exposure to human immunodeficiency virus (HIV), the virus that causes AIDS [18]. Resistance was traced to a small protein, known as the *CCR5* chemokine receptor, which is on the surface of certain immune cells. This protein is ordinarily the portal by which the AIDS virus gains entry into immune cells and thus initiates a process that kills immune cells. Yet certain people have a mutation in the *CCR5* gene that blocks expression of the receptor protein and thus makes them resistant to AIDS. This mutation – called the CCR5-Δ32 mutation – may have arisen as recently as 700 years ago. It has been proposed that the Δ32 mutation then underwent very strong selective pressure such that it increased in frequency among ancestral Europeans. But what could this selective pressure have been?

The selective pressure on the CCR5-Δ32 mutation may have been the Black Death, a devastating plague that swept through Europe over and over again from 1347 until 1670 [19]. The Black Death – so called because of subcutaneous hemorrhages that formed black splotches on the skin of victims – first arrived at the

Sicilian port of Messina in 1347 and moved inexorably northward, reaching the Arctic Circle by 1350. Everywhere it went it laid waste, killing 40% of the people across Europe and as many as 80% of the people in cities such as Florence before temporarily burning itself out in 1352. While it is impossible to prove this now, some scientists believe that the CCR5-Δ32 mutation may have conferred some resistance to the Black Death, perhaps by the same mechanism as for AIDS, by blocking access of the plague bacterium to immune cells. Clearly, some people in the Middle Ages must have been resistant to the Black Death; one monk in a monastery survived after ministering to and burying every one of his brothers in faith. The carnage wrought by the Black Death forced sweeping demographic changes upon a stunned and mourning Europe, including the end of feudalism, but it also may have made modern people more resistant to AIDS. Today, in parts of the northern Europe, the prevalence of the CCR5-Δ32 mutation is 10%, but it can be as high as 18%. Thus, the scope and scale of the Black Death in the 14th Century potentially made the AIDS epidemic in the 20th Century less devastating than it could have been.

As appealing as this story is, it is highly contentious. There is recent evidence that the CCR5-Δ32 mutation is much more ancient than 700 years old; DNA from bones dated to the Bronze Age has shown that this mutation may have been present more than 3,000 years ago [20]. This weakens the argument that the CCR5-Δ32 mutation was selected by the Black Death. Another problem is that the Black Death did not strike Europe alone; it may have killed more people in China, North Africa, and the Middle East, yet the CCR5-Δ32 mutation is not found outside of Europe [21]. This might be just an evolutionary accident; if a mutation does not arise, then it cannot be selected for. Even within Europe, prevalence of the mutation is opposite to what would be predicted based on plague mortality. If the Black Death truly selected for the CCR5-Δ32 mutation, then Sweden which has the highest prevalence of the mutation, should have been hardest hit by the plague, whereas Greece and Italy, where the mutation is rare, should have been spared. Yet historical records show that the Black Death killed far more people in southern Europe, near the Mediterranean, than in northern Europe. A final compelling fact argues against the idea that the CCR5-Δ32 mutation protects against the Black Death. Bones from a mass grave site in Lubeck, Germany, which contains victims of the plague of 1348, show the same frequency of the CCR5-Δ32 mutation that is seen in bones from people buried before the plague [22]. If the CCR5-Δ32 mutation had conferred even partial resistance to the plague, one would expect the mutation frequency to be lower among those who died of plague. In effect, the evidence argues strongly that there is little or no relationship between the gene that provides resistance to AIDS and any genes that might have provided resistance to the Black Death. And, even though the CCR5-Δ32 mutation may have reached high prevalence in northern Europe rather quickly – in an evolutionary sense – it still took at least 3,000 years to do so.

Another example of rapid evolution of human genes under what may have been stringent selection is provided by the lactase gene [23]. In many people, the ability to digest lactose – the sugar common in milk – disappears in childhood, but in Europeans lactase activity often persists into adulthood. This may have provided an evolutionary advantage, as it would have enabled cold-climate Europeans to rely less on farming while obtaining a significant portion of their diet from dairy production. Scientists sequenced genes from 101 Europeans, in an effort to determine how much variability exists in the relevant genetic sequence. This type of information could enable scientists to back-calculate how long ago the mutation arose, assuming that the mutation rate is constant for all people. This work suggests that the lactose tolerance arose quite recently and may have provided a strong selective advantage to those people who retained an ability to digest lactose. Nevertheless, calculation suggests that lactose tolerance arose between 5,000 and 10,000 years ago, meaning that evolutionary change is quite slow even under strong selection.

In short, evolution cannot possibly explain an increase of 5 IQ points per decade. Perhaps the strongest evidence that evolution is inadequate to explain increasing intelligence comes from a study of intelligence itself. A study performed in the Netherlands about 25 years ago [24] was never published, but has been described in detail in a review article by James Flynn, the man for whom the Flynn effect is named [25]. The original study examined a sample of 2,847 men who, at the age of 18 years in 1982, took a test that is given to all Dutch men, to assess their strengths and abilities prior to induction into the military. The same test has been given to men in the Netherlands since 1945, so it was possible to find test results for the fathers of all of the men who took the test in 1982. On average, the fathers had taken the test 28 years earlier than their sons. Not surprisingly, there was a fairly good correlation between the IQ scores of fathers and sons. Yet the sons scored 18 IQ points higher than their fathers, which amount to a 6 point IQ change per decade [24]. Because this enormous change in IQ took place over a single generation, it is not possible for evolution to have had an effect. Absent a massive mortality among the generation of men tested in 1982, which we know did not happen, there was no time for natural selection to have mediated *any* change in gene frequency in the population.

## Human Brain Evolution Is Recent and Rapid

We have argued that there is no evidence to support the idea that the Flynn effect results from evolution. Nevertheless, there is strong evidence that the evolution of the human brain has been recent and rapid, albeit on a slower time scale than needed to explain the Flynn effect.

One of the most easily observed traits of the human brain is its size relative to the brain of chimpanzees and other primates. Our brain is at least three times as large as the chimpanzee brain, our nearest living relative from which we diverged 7–8 million years ago. Human brain size is regulated by at least six different genes, and mutation of any of these genes causes microcephaly, a medical condition in which the brain is dramatically smaller than normal [26]. In patients with microcephaly, reduction in the volume of the brain is combined with severe mental retardation, even though brain structure is apparently normal and there are few effects elsewhere in the body. If a patient has a mutation of one *microcephalin* (or *MCPH*) gene, brain volume may be reduced to 400 cubic centimeters (cc), compared to a normal volume of nearly 1,400 cc. Thus, mutation of an *MCPH* gene leads to a 70% reduction in the total brain volume. This gene is thought to control the rate of division of neural stem cells during the formation of the nervous system, though this is not known for certain.

A study of DNA samples from 86 people around the world suggests that one particular form of the gene, called *MCPH1*, is more common than all others [26]. Abundance of this gene is far higher than would be predicted by chance alone, meaning that it may be an adaptive mutation that has undergone strong selective pressure. Using the logic that the "molecular clock" ticks at a constant rate, such that stepwise changes in DNA can give insight into how long ago a mutation arose, it is possible to estimate when this mutation first occurred. Analysis suggests that the *MCPH1* mutation arose in humans only 37,000 years ago, even though the human genome is at least 1.7 million years old. Interestingly, the date of this mutation roughly coincides with the time period during which Europe was colonized by Cro-Magnons emigrating from Africa. Although we cannot be certain that *MCPH1* increased in frequency due to natural selection for brain size, it is certain that natural selection has acted upon the human brain rather recently.

Another study confirms and extends the study of *MCPH1*, reporting that a second gene has also undergone recent and rapid selection in humans [27]. This gene, known as *ASPM* or *MCPH5*, is also one of the six genes that can produce human microcephaly. The *ASPM* gene was sequenced in the same 86 people, and again it was found that one variant was more common than any other. The molecular clock argument suggests that this variant appeared a mere 5,800 years ago, after which it underwent a very rapid increase. If the date of origin of this new gene variant is correct, this implies that the human brain is undergoing extremely rapid evolution, perhaps due to selection for increasing brain volume. Nevertheless, this rate of evolution is still far too slow to explain a 5 point rise in IQ per decade.

There is now a long and growing list of genes that may contribute to the development of the human brain [28]. Yet the best evidence suggests that it is highly unlikely that any of these genes has undergone a significant change in prevalence at a rate rapid enough to explain the Flynn effect.

# Chapter 4
# Brain Development and Increasing Intelligence

In the past, it was essentially impossible to study brain changes in a healthy child because no means existed to examine the brain without doing at least some harm to the child. Brain surgery is only done as a last resort, when someone is desperately ill, hence, this option was ruled out as a way to understand the developing brain. Autopsy is permissible in a child who has died, but there is always a concern whether a child who has died is fundamentally different from a child who is well. Medical imaging could have been done, but medical imaging carried some risk for the person being imaged. Over 100 years ago, X-rays were discovered, so they could have been used to study the developing brain, but this was not done because of the risks from radiation; even low levels of radiation can be harmful to the growing brain [1]. More recently, computed tomography (CT) became available, which yields far more detailed images, but CT still requires radiation exposure, so the benefits of research do not outweigh the risks of harm [2].

Recent advances in medical imaging have reduced the radiation-related risk concerns, making it safe to study the developing brain. Using a method called magnetic resonance imaging or MRI, it is possible to visualize the brain at every phase of its growth, from the fetus to the fully adult, without exposing a person to harmful radiation. The MRI method is too complex to be explained in detail, but the images are simply maps of the distribution and abundance of water in brain tissue. Because the brain is soft tissue containing a great deal of water, these "water maps" form detailed and genuinely beautiful images that enable clinicians to visualize the brain with clear and compelling detail. Contrast between structures in the image is largely a function of the water content of brain tissue. Thus, a bright area in an image can reveal brain regions with a great deal of free water, such as the fluid-filled spaces within or around the brain. Conversely, dark areas in an image can reveal brain regions that have relatively little water content, such as the cortical gray matter. These images are so detailed and so faithful to brain anatomy that neurosurgeons routinely use them while planning surgery.

Brain MRI has now been used to study hundreds of healthy children, to understand how the brain grows and develops. As a result of these studies, we now know that there is a complex pattern of developmental change in the brain as children

R.G. Steen, *Human Intelligence and Medical Illness*, The Springer Series
on Human Exceptionality, DOI 10.1007/978-1-4419-0092-0_4,
© Springer Science+Business Media, LLC 2009

mature, and that these physical changes correlate with cognitive and emotional maturation. But these changes do not appear to explain the Flynn effect.

## Patterns of Brain Growth and Development

A striking and very important trend observed in human brain development is that the final size of the brain relative to the body is enormous, when compared to our closest primate relatives. This is the conclusion drawn by scientists [3] who visualized the brain by MRI in humans and in 10 primate species (e.g., rhesus monkeys, baboons, gibbons, orangutans, chimpanzees, gorillas). Analysis, based on trends among the rest of the primates, showed that the human brain is far larger than expected. If brain volume is plotted as a function of body weight, the primates form a reasonably clear relationship, with larger primates having larger brains. But the human brain is at least threefold larger than predicted from the relationship between brain and body size in other primates. For example, orangutans are somewhat larger than humans in body weight, but their brains are 69% smaller than the human brain. The overall increase in human brain volume is driven, in part, by an increase in volume specifically of the forebrain, directly above the eyes. This suggests that there has been a rapid evolution, specifically of that part of the brain that controls language and social intelligence [4].

Another striking difference between the human and the primate brain is that the human brain is more deeply folded or gyrified than expected, compared to primates [3]. There has been a great deal of speculation regarding the importance of gyrification, but most scientists agree it is important; in humans an abnormality called lissencephaly – in which the brain is largely free of gyres – is associated with profound mental retardation. There is speculation that a highly gyrified brain may be a cognitive advantage, in that the deep folds provide a larger surface area for a given volume. Because cortical gray matter – the "thinking" part of the brain – is at the brain surface, a heavily-gyrified brain can have an increased amount of cortex relative to the volume of the rest of the brain. Another potential benefit of cortical folding is that gyrification might enhance the speed of processing in the brain. This is because white matter – the "wiring" of the brain – can more easily form connections between distant parts of the brain if some of the connecting "wires" are able to bypass parts of the brain that are not relevant to the task at hand. In other words, if unwanted cortex is located on a fold, then the wiring between necessary parts of the brain can cut across the fold and thereby traverse a shorter distance. Such shortened point-to-point connections could potentially result in faster information processing in the brain.

A third difference between human and primate brains is that the volume of white matter in particular is increased in the human brain [3]. Considering that the white matter is merely "wiring", it might seem trivial that the human brain has more wiring than other primates. But this difference is probably not trivial at all; if there is more white matter, this suggests that the number of connections between neurons has increased faster than the number of neurons during the evolutionary process. Without dense connections between distant neurons, it may not have been possible for the human brain to develop task specialization or intelligence or even consciousness [4].

If one evaluates the pattern of human brain growth, several additional surprises emerge. For example, the human brain grows very rapidly compared to the rest of the body. Human brain volume increases roughly threefold from birth to age 5 [5] but there is only a 10% increase in brain volume thereafter [6]. The average child of age 12 has a brain that is fully adult in volume [7], even though most 12 year olds have a far smaller body size than an adult. Volumetric growth of the brain is accelerated with respect to the rest of the body, so that brain volume is adult-like well before the body attains a mature size.

The fact that the child's brain grows rapidly in volume does not mean that the average 12 year old has an adult brain. While any parent of a teenager would have guessed this, science can now describe in detail changes that happen in the maturing brain, well after the brain has already achieved adult volume. The volume of gray matter is at a maximum sometime between 6 [8] and 12 years of age [9], then the gray matter begins to decrease in volume. Many studies have reported that the gray matter volume decreases with age in adults [10], but the fact that the gray matter volume also decreases with age in both adolescents [11, 12] and children [13, 14] is something of a surprise. These results imply that gray matter begins to atrophy during adolescence. Alternatively, results could mean that gray matter volume is lost in some other way.

Changes in gray matter volume seem to be linked to changes in white matter volume. It has been proposed that some brain areas that are visualized as gray matter by MRI in adolescents will undergo maturational change, and eventually become white matter [15]. This is suggested by the fact that the volume of white matter continues to increase well past the age of 20 [16], perhaps to as late as age 50 [16]. White matter volume growth is a result of ongoing myelination – the process in which an insulating layer literally grows around the neuronal "wires" connecting one part of the brain to another. Myelin substantially increases the speed of conduction of nerve impulses down the length of a neuron, so myelination probably increases the rate of information processing in the brain. In essence, myelin forms sheaths around conducting "wires" in the brain much the way a plastic sheath insulates the wires of a computer.

The infant brain is very poorly myelinated compared to the adult brain, but myelination occurs at an astonishing rate during infancy and childhood [17]. However, there can be a great deal of variation in the rate of myelination from one brain region to another. Autopsy studies show that microscopic spots of myelin are present in some infants in the posterior frontal white matter (above and behind the eyes) as early as 39 weeks post-conception, about one week before birth is likely to occur. By week 47, just 7 weeks after birth, about half of all infants begin to myelinate the posterior frontal white matter, and myelination is more or less complete by week 139, insofar as an autopsy can reveal. In contrast, in the sub-cortical association fibers of the temporal lobe, most infants do not begin to myelinate until week 72. But, even though it starts late, myelination is completed in the sub-cortical association fibers at about the same time as in the frontal white matter. Thus, in different brain regions, white matter myelination begins at different times, and proceeds at differing rates, but reaches completion more or less concurrently.

## Brain Myelination and Developmental Maturity

There are several general patterns of myelination in the infant brain [18]. The central part of the brain myelinates before the periphery, which probably reflects the fact that the central part of the brain is concerned with very basic functions – moving and sensing the environment – without which life would not be possible. After these functions mature and can be performed reliably, it becomes possible to develop more advanced skills. Sensory pathways mature before motor pathways, central connections are established before peripheral connections, and the control functions of the brain are strengthened before real thought can happen. In other words, myelination proceeds by a sort of "hierarchy of need." After the most important functions are established, the brain can begin to elaborate those functions that are less essential but more characteristic of a mature brain [19]. This reasoning assumes that a nerve tract is not fully functional until it is well myelinated, which may not really be true. Yet most neuroscientists would agree that the time between when myelination begins and when it is completed is a period of vulnerability for a child. Any brain injury that occurs during such a developmental "window" could interfere with the adult function.

There can also be some variation in the rate of myelination between infants, as some infants apparently are able to myelinate white matter sooner than others [20]. This makes a degree of sense, since it is certainly true that developmental milestones are reached by different infants at different ages. Recent studies using a novel MRI method called diffusion-tensor imaging (DTI) suggest that white matter myelination takes years to reach completion, and that the functionality of neuronal tracts continues to change long after the initial layers of myelin have been laid down. There are apparently maturational processes that affect the integrity of white matter tracts, hence some parts of the brain may only be able to interact with other parts after certain developmental milestones have been reached. In fact, it could be that developmental milestones in infants (e.g., sitting, walking, talking) occur only because some particular brain structure has matured.

Direct evidence linking myelination to behavior may be at hand, in the form of a study that correlated white matter maturation with reading ability [21]. Most children learn to read rather quickly, but roughly 10% of children have problems reading that cannot be explained by poor schooling, lack of intelligence, or inadequate opportunity. A group of 32 children of varying reading skills was evaluated by DTI, to assess white matter maturation. As white matter tracts become more adult-like in structure, the DTI signal changes in a predictable way. Immature white matter shows a relatively loose and incoherent structure, whereas mature white matter is tightly-packed and highly organized. Because these two different forms of myelin look quite different by DTI, any disturbance in myelination is easily visualized, provided all subjects are of the same age. Children in this study were imaged by DTI and were tested for reading skill, using a word identification test. A strong correlation was found between reading ability and the degree of maturity in a specific white matter tract in the brain. The structure of white matter in this region did not correlate well with age or with non-verbal intelligence, but it did correlate with reading ability. In fact, variation in the DTI signal was able to explain about 29% of the variation in reading ability, which was statistically significant and could well be clinically significant. Therefore, white matter maturation appears to play a role in the acquisition of reading ability. Logic would suggest that maturation of white matter may explain much about the age-related acquisition of ability in humans.

Brain myelination apparently continues throughout adolescence and well into adulthood. At age 9, the average white matter volume is about 85% of the adult volume, and white matter does not reach a maximum volume until age 40–48 years [16, 22]. Gray matter volume decreases with the increase in white matter volume, such that there are rapid changes in the relative proportion of gray matter to white matter until about age 25 [15]. Because the overall brain volume is stable by age 12 [7], this implies that any decrease in gray matter volume after age 12 must be offset by an increase in the volume of either white matter [15] or the cerebrospinal fluid around the brain. Whether the decrease in gray matter volume between age 20 and age 40 can be entirely explained by white matter myelination [11], or whether there is also a component of gray matter atrophy involved [23], remains unknown. Overall, brain volume certainly begins to decrease by the sixth decade of life as a result of tissue atrophy, though there does not seem to be an associated cognitive decline until perhaps as long as two decades after atrophy begins.

It is likely that there are also maturational processes in gray matter, which may be similar in principle to the maturation of white matter during myelination. Perhaps the loss in volume of gray matter in adolescence and adulthood is not only a function of atrophy; maturing gray matter may undergo a process of compaction so that the same neurons fill a smaller volume [24]. It is possible that the brain is not fully adult until after gray matter maturation has occurred, though this is speculative at present. No methods have yet been developed that are able to visualize brain volume changes with a sufficient degree of precision and accuracy to know whether there is a maturational process in gray matter [25]. Yet there is evidence that intelligence is associated with the development of the cortex, and that the schedule of brain maturation has a huge impact on the development of intelligence [26].

It is a reasonable hypothesis that mature brain function is not possible until the structure that supports a function is mature. For example, children are impulsive, prone to heedless action, and likely to commit violence, probably because the frontal lobes of their brains have not yet matured [27]. The frontal lobes of the brain give us the ability to restrain our actions, to use foresight and planning, to edit our impulses. The incompletely myelinated frontal lobes that are characteristic of adolescence may simply be unable to perform such executive functions, which could explain a great deal about adolescent behavior. If this hypothesis proves true, it has a crucial implication: adolescents who cannot exercise mature executive control over themselves should not be held criminally culpable to the same degree as an adult. Juveniles have a diminished capacity to form intent compared to an adult, and they probably have a diminished capacity to restrain themselves from committing violence. In short, adolescents cannot be expected to act like adults since they lack the tools to do so.

## Is Education Now Better Able to Compensate for Differences in Developmental Maturity?

One potential way to explain the Flynn effect is that educators have learned to work within the limitations of the developing brain more effectively. Perhaps, as our understanding of brain maturation increased, we have learned to better capitalize on our strengths and minimize our weaknesses. In short, education may simply be more efficient now than it was in the past. In essence, this is a variant of the hypothesis that IQ tests measure achievement rather than aptitude, since superior teaching should not be able to impact aptitude, though good teaching can certainly increase achievement.

An interesting test of the hypothesis of increasing educational efficacy has been provided by a study in France, which evaluated the ability of young children to reason [28]. The psychologist, Jean Piaget, studied the cognitive development of infants, using a series of very clever tasks that test the ability of infants to gain insight into the underlying nature of a problem. Such "Piagetean" tasks have been adapted and used to test young children, and it is known that strong performance on a Piagetean test is associated with high IQ in traditional tests of intelligence. Yet Piagetean tests differ from ordinary IQ tests in several important ways. First, Piagetean tests do not have time limits, thus success depends upon accuracy rather than speed. This is a key distinction, because some scientists have sought to explain the Flynn effect as a result of faster cognitive processing in modern children. Second, Piagetean tests require that a child be able to explain why they made a particular choice. This eliminates random chance or informed guessing as a consideration, since a guess that cannot be defended is not accepted as correct. A drawback of this kind of test is that it may mean that Piagetean tasks are rather more weighted toward verbal skills than are Raven's Matrices, where an answer is accepted as correct whether or not it can be defended. Yet the requirement that children explain their answer is particularly valuable in assessing whether a child has insight into a problem; since tasks are designed so that insight is required, this means that Piagetean tasks are strongly weighted to test reasoning ability. Third, Piagetean tasks have been validated with a variety of methods, so that one can be sure that a high score in such a test is actually meaningful. Finally, Piagetean tasks are designed to assess facility with knowledge that can be important in understanding how the world works.

A key feature of Piagetean tests is that they use thinking patterns and skills that are unlikely to have been learned in school. Someone who performs well in a Piagetean test therefore shows reasoning skill, rather than skill in manipulating concepts learned in school. For example, one Piagetean test involves an understanding of the conservation of mass [28]. An adolescent is shown three balls of identical size – one made of metal and two made of clay – together with two containers of water. The first task is to dissociate weight from volume, to realize that the metal ball and the clay ball displace the same volume of water, though the balls have different weights. So, for example, the test subject would be asked why the water level rises when a ball is dropped into the container of water, then asked to predict the height of water in a container when a heavier ball of the same volume is dropped in the water. The accepted answer is that the balls, being of equal volume, displace equal volumes of water. The second task is for the adolescent to realize that volume is conserved, even if the shape is changed. In this task, the examiner manipulates one of the clay balls, rolling it into an elongated shape or cutting it into segments, to see whether the adolescent understands that the volume of clay has not changed, no matter what shape it assumes. The final task is for the test subject to realize that weight is conserved, even if the shape changes. In this task, one ball of clay is compared to the other, with the second ball of clay being flattened or cut into pieces while the adolescent watches. This task, which is conceptually a bit simpler than the other tasks, is used only if a subject has done poorly on the preceding tasks; if an adolescent has done well in the other tasks, this last task is skipped. There are other Piagetean tasks, testing an understanding of how objects are arranged in combination, how simple probability governs what happens, how a pendulum oscillates, and how curves are drawn by a simple machine. Each task involves simple tools used in simple ways, but ways that the test subject has probably never seen before. Each task involves an insight into the way that things work and an explanation of that insight. But these simple tasks enable the tester to understand how a person makes sense of the things that they see.

A group of 90 adolescents between the ages of 10 and 12 years was tested in 1993, and the results of these tests were compared to a normative database assembled in 1972 [28]. The adolescents tested in 1993 scored substantially better than the normative sample, achieving scores equivalent to a 3.5-point increase in IQ over 21 years. In a second study, 90 adolescents between the ages of 13 and 15 years were compared to adolescents first tested in 1967. In the second study, modern adolescents also scored substantially better than the normative sample. It is somewhat risky to impute an IQ score for a test that is scored in a different way than a true IQ test, but it is clear that test performance improved substantially over a rather short period of time. What is especially striking is that adolescents in the normative sample and in the later sample tended to use the same strategies to solve the puzzles, but the later group of children was more successful. Because a child is unlikely to have learned how to solve such puzzles in school, it seems unlikely that education could explain the increase in scores on Piagetean tasks. These findings strongly suggest that education, though it may be more effective now than in the past, cannot explain the Flynn effect.

Yet things are never as simple as we would like them to be. It is impossible to completely dismiss education as a cause of rising IQ scores, because children are spending much more time in the school now than in the recent past [29]. Since about 1980, the high school graduation rate in the United States has been steady at around 85%, but in 1950 the high school graduation rate was less than 35%. Similarly, in 2003, 28% of adults had a bachelor's degree from college, but in 1950 only about 6% of adults had a bachelor's degree. Even if education is relatively ineffective, it could be that the sheer amount of time spent in the classroom has a positive effect on the IQ scores of American adolescents. Yet this assumes that IQ tests are tests of achievement, not tests of aptitude. If IQ tests are indeed tests of aptitude, one would not expect school attendance to make a difference. This is especially true if IQ is assessed with untimed (Piagetean) reasoning skills [28] or if it is assessed with a non-verbal test like Raven's Matrices.

Even if increasing time in school explains the rising IQ scores in American adolescents, educational effectiveness cannot explain why IQ scores are also rising in young children. For example, in the study of children in rural Kenya [30], IQ tests were given within 4 months of the children entering school, well before school attendance could have had an impact. Furthermore, testing used Raven's Matrices, and Kenyan teachers are unlikely to have been able to "coach" children to perform well on Raven's, since the teachers themselves were unfamiliar with the test. Finally, Raven's Matrices is a non-verbal test of reasoning ability; schooling has an enormous impact on verbal ability, but a lesser impact on reasoning ability. In short, the Kenyan study suggests that, even if teachers are more effective, this cannot explain rising IQ scores.

It is nonetheless possible that children benefit by having parents who attended school, that the Flynn effect is a reflection of cumulative education of the family. But even this possibility seems unable to explain what happened in rural Kenya [30]. In 1984, 26% of Kenyan mothers reported having had no school at all and only 7% reported having a "Standard 8" education. By contrast, in 1998, after children's IQ scores had increased 26 points, 9% of mothers still reported having had no schooling and 18% reported having had a "Standard 8" education. Thus, while the proportion of parents with some schooling had increased, the overall levels of school attendance were still low, making it unlikely that school attendance had a major impact on the children.

## Is Increasing Environmental Complexity Producing a Rise in IQ?

It has been argued that environmental complexity is sharply increased in the modern world, and that children benefit cognitively from the richness of stimuli that surrounds them. In 2005, a book entitled *Everything Bad is Good for You* even argued that a steady diet of television and video games has a beneficial effect on the cognitive development of children [31]. This is an appealing idea for many reasons.

First, it comforts stressed parents – too busy with work to entertain their children – and reassures them about choices the parents made in raising their children. Secondly, it acknowledges that environmental complexity has a powerful effect in experimental animals, by increasing the rate of birth of new neurons in the brain. Finally, it makes intuitive sense that video games can enhance cognitive maturation; snap decisions with inadequate data in a highly fluid environment are required for success in the video environment as in life itself, so video games may be a place to learn life skills without getting life bruises.

Yet it is highly unlikely that environmental complexity can explain rising IQ in children from rural Kenya [30]. Store-bought toys, video games, television, even colorfully-printed cereal boxes were essentially unknown in 1984, when the study began, and they were still a rarity in 1998, when the study ended. At the start of the study, no family had a television and at the end of the study, only 9% of families had a television. Nursery school attendance was just 7% in 1984, and it was available to only 15% of children in 1998. Sunday school attendance had been 90% in 1984, and it was 99% in 1998, so even this did not change much. By every available measure, the environment was not much more complicated in 1998 than it was in 1984, yet IQ scores rose by 26 points over this time period.

The great weakness of the "environmental complexity" hypothesis is that it has never been tested in people. Experimental studies of environmental richness use rats or mice in a laboratory environment, which is almost certainly stark and uncomplicated by comparison to what a rodent would experience in nature [4]. What this may mean is that stimulus enrichment can augment cognition in animals – or perhaps even in people – but only among those who live in a severely depauperate environment. Stimulus enrichment may have little or no impact in an environment of "normal" complexity. However we cannot know this for certain; it is unethical to raise children in an environment deliberately made depauperate, since anecdotal reports suggest that such simple environments can be very harmful to children [4].

## Hypothesis: Child Development is Happening Sooner or Faster than in the Past

A potential explanation for steadily increasing intelligence in people is that brain maturation is happening faster now than in our forebears. Normally, a child's performance on an IQ test is compared to the performance of a normative sample of children, some of whom were tested at the same chronological age. Thus, if child development is simply happening faster now than in the past, children will be intellectually more mature at the same chronological age. If a modern child is compared to a child in the normative data base, the modern child might be expected to have an older intellectual age. For example, a child who is 12 years and 8 months old now may be intellectually comparable to a child who was 13 years and 4 months old when normative data were gathered 20 years ago. Yet this difference in maturation rate may have little or no practical importance if both children grew up to have the same adult IQ. Nevertheless, if maturation rates are faster now, this would make it seem that children are smarter now than they were in the past.

Is there any evidence that cognitive maturation can vary in a way that impacts measured IQ? While there is no definitive proof yet, a fascinating study conducted in Estonia suggests that there can be meaningful variation in the rate of cognitive maturation [32]. Raven's Matrices, the non-verbal test of reasoning ability, were used to test nearly 5,000 school children in Estonia, ranging in age from 7 to 19 years. When the IQ of Estonian children was compared to children in Britain and Iceland, who comprised the normative database, it was found that the youngest Estonian children scored better than the normative sample. However, after first grade, the Estonian children fell behind the normative sample and remained behind until age 12. After age 12, the average IQ of Estonian children again surpassed the normative sample. Scores from a large number of children were evaluated, which probably precludes random variation or accrual bias as a reason for the variation observed. The simplest explanation for these results may be that Estonian children simply follow a different developmental trajectory than do children in Britain or Iceland. The cognitive development of Estonian children appears to proceed somewhat slower than "normal" during elementary school, but then rebound and overtake the cognitive development of other children after about Grade 6. This explanation seems plausible because immigration to Estonia was quite low until recently, so Estonian children are genetically rather uniform, and perhaps more likely to mature at a similar rate. These results suggest that cognitive maturation can take longer in some children than in others, and that this can lead to meaningful variations in group IQ. This is also consistent with a hypothesis that cognitive maturation can happen faster than expected, which could lead to an apparent increase in IQ in some children.

This fairly simple hypothesis –called the "fast maturation" hypothesis – is rather hard to test in practice. Those data that are available do not go back in time very far, and most of the data tend to be imprecise anyway. Imprecise data can, of course, be used in a study if a great many subjects are involved, since minor and meaningless variations in a measurement can be overcome by the brute-force approach of having a huge number of subjects. Yet studies that are relevant to testing the "fast maturation" hypothesis tend to be small as well as recent.

## The Timing of Puberty in Adolescence

Perhaps the best test available of the "fast maturation" hypothesis is provided by observations on the timing of puberty among adolescents. It has been the subjective impression for years, among pediatricians in the United States and in Europe, that girls enter puberty at a younger age now than in the past [33]. There are charts available to pediatricians that detail the "normal" age at which secondary sexual characteristics develop, including age at development of breasts, growth of armpit and pubic hair, or first menstruation ("menarche"). Such signs of maturation are unequivocal, and pediatricians had noticed that more and more of their young patients are undergoing precocious puberty, compared to the norms in the old published charts.

To update these charts, a massive effort was undertaken by the Pediatric Research in Office Settings (PROS) network, a professional alliance within the American Academy of Pediatrics. Between 1992 and 1993, physicians from all over the United States collaborated in examining 17,077 girls, with the racial balance of the study sample reflecting the racial balance of the nation as a whole [34]. Strikingly, among the 17,077 children examined, about 1% of girls showed breast or pubic hair development at an age of only 3 years. By the age of 8 years, 15% of girls had begun to show signs of puberty. Previously, the prevailing wisdom was that only 1% of girls would show signs of puberty by age 8. The average age of breast development in girls in the PROS study was 10.0 years, the average age for pubic hair development was 10.5 years, and the average age of menarche was 12.9 years. Thus, girls were beginning to mature sexually about 6–12 months sooner than expected, based on previous studies. However, though the age at breast and pubic hair development was younger than in the past, the age at menarche was not substantially different.

These findings could perhaps be the result of an accrual bias; if parents were aware of early pubertal changes in their daughters and if they brought children to the doctor specifically because of a concern that puberty was happening "too soon," these children would not be representative of the nation at large. In other words, concern about the development of breasts or pubic hair may have been a "hidden agenda" for parents. Yet it is likely that parents concerned about early puberty were offset by parents concerned about late puberty, so this study may not have had a significant accrual bias. Furthermore, even if an accrual bias was present, this study sample is still representative of children brought to a pediatrician. Interestingly, girls in this study were also taller and heavier than girls in the earlier study that had been done in the United States. This suggests that the precocity of puberty may reflect a general precocity of development among children in the United States, which could be an important finding.

There is a clear trend over the past few decades for increasing obesity in American children, and nutritional status can affect pubertal timing [35]. Hence, it is possible that precocious puberty may be associated with obesity in young girls. To test this hypothesis, the PROS data were reevaluated, comparing the body mass index (BMI) of girls showing signs of puberty with the BMI of girls who were not pubertal. This clever reanalysis of data revealed that pubertal girls were significantly more likely to be obese than were non-pubertal girls. Even if every possible confounder was controlled statistically, obesity was a significant contributor to early puberty. These results suggest a key conclusion; child development is critically dependent upon nutrition. One would expect, therefore, that inadequate nutrition would delay physical development – and could potentially interfere with cognitive development. As can be seen, these ideas have far-reaching implications.

Results of the PROS study have now been confirmed. Researchers used data from the National Health and Nutrition Examination Survey III (NHANES III), which was conducted in the United States between 1988 and 1994, to assess the age of puberty in both girls and boys [36]. This survey may be more representative of the nation as a whole, though it enrolled only 2,145 girls. Based on the NHANES III data, the average age of early pubertal change in girls was essentially the same as in the PROS study, and considerably younger than the old reference charts would have it. Unfortunately, the NHANES data did not assess menarche, so what may be the most critical piece of information is missing.

A strength of the NHANES study was that it offered evidence that boys were also reaching puberty at a younger age than in the past [36]. This would be expected if puberty is related to nutritional sufficiency, since nutrition is as much an issue in boys as it is in girls. In fact, the finding that pubertal precocity is present in boys as well as in girls would seem to disprove one of the more fashionable but less plausible ideas. Environmental estrogens have been blamed for early puberty in girls, but estrogens cannot explain why boys are also attaining puberty sooner.

The age at which girls entered puberty in the past remains somewhat controversial [33]. Those studies upon which the old reference charts were based were often small and the samples were not necessarily representative of the nation as a whole, so it is hard to know if the accepted values were actually good estimates of age at puberty. Sometimes the way in which puberty was assessed in the old studies was not clearly spelled out, so it is even possible that some of those studies were done differently than they would be done today. Nevertheless, there is broad acceptance that the secondary signs of puberty happen earlier now than in the past, though the age at menarche probably has not changed much, if at all. There is also broad acceptance of the finding that early-maturing girls are likely to be obese – whether measured by BMI or skinfold thickness or percent body fat – compared to late-maturing girls. Few would argue against the idea that BMI correlates better with maturational age than with chronological age, and that early-maturing girls tend to be more obese in adulthood than late-maturing girls. But whether obesity in some way causes early onset of puberty is more controversial.

Is it reasonable that the "fast maturation" seen in pubertal timing could also have an impact on cognitive development? It is noteworthy that modern children are taller now than in the past, which suggests that diet is more likely to be adequate now. If brain maturation was in any way limited by diet in the past, then diet should be of less concern anywhere the average height of children has increased. It is an ironic possibility that a surfeit of nutrients is bad for the body, because it increases the risk of diabetes, heart disease, stroke, and cancer, while it may perhaps be beneficial for the brain, if some nutrients that were once limiting are no longer as limiting.

These considerations raise an intriguing hypothesis, noteworthy for its simplicity. Perhaps modern children are simply healthier now – even considering obesity – than in the past. It would be expected that healthier children would be better able to concentrate at school, better able to focus during tests, better able to sustain attention, better able to concentrate, better able to react quickly. Could it be that IQ is simply a surrogate measure of health? Is it reasonable to suppose that increasing health could explain rising IQ throughout the world?

# Chapter 5
# Environment and Increasing Intelligence

Rising IQ certainly cannot be attributed to genetic change; evolution is a far slower process than that. If evolution alone were responsible for increasing intelligence, it is likely that the rate of IQ increase would be less than 1% of what it is now. In other words, if evolution were the driving force, we would not expect IQ to increase by a point every 2 years; we might expect it to increase by a point every 200 years. Clearly, there must be another reason for increasing intelligence.

We have spent some effort in the last few chapters showing that the increase in IQ score also cannot be explained by flaws in the tests or in the way that the tests are administered. Neither is rising IQ a function of accelerated child development, since it is not clear whether development actually is happening any faster than it was before. What is left, if we are to explain the Flynn effect? The only remaining possibility seems to be an answer that is almost too simple to be true; the environment is changing in a way that enables IQ to rise.

## Hypothesis: The Family Environment is Improving, Thereby Enabling Intellectual Growth

Is the family environment now more supportive of children than it has been in the past? Is it possible that modern families are better able to foster the cognitive growth of their children than in past generations? Are more families intact and more children able to develop their inherent potential? Did, for example, the increase in employment rate of parents under the Clinton Administration enable more parents to have sufficient resources for their children? Or does the decrease in teen pregnancy mean that fewer children are being raised by mothers overwhelmed by poverty? To what extent can rising IQ be attributed to improvements in the family and social environment of children?

R.G. Steen, *Human Intelligence and Medical Illness*, The Springer Series
on Human Exceptionality, DOI 10.1007/978-1-4419-0092-0_5,
© Springer Science+Business Media, LLC 2009

## Is the Social Environment Contributing to the Rise in IQ?

According to the testimony provided by Ron Haskins, a Senior Fellow at the Center on Children and Families at the Brookings Institution, given to the Senate Committee on Appropriations in 2006, there is no good news for children in terms of the social environment [1]. According to Haskins, "children do best when raised by their married parents," yet the marriage rate for parents has been falling for decades. At the same time, the divorce rate has risen sharply and the percentage of births to unmarried women has soared. The percentage of children in single-parent families has increased relentlessly to a high of 28% in 2004. Poverty in female-headed households is almost fivefold higher than poverty in married-couple households, and poverty has a profound effect on educational attainment and school performance. According to Haskins:

> "Mothers who give birth outside marriage are also more likely to be high school dropouts, to live in poverty, and to be unemployed, all of which are correlated with poor developmental outcomes for children… [T]he percentage of babies born outside marriage rose from under 5 percent in the 1950s to about 33 percent in 1995 before falling for the first time in decades… In 2000, for example, the share of babies born outside marriage for whites, Hispanics, and African Americans were 22 percent, 43 percent, and 69 percent respectively." [1]

Another pessimistic view of the impact of the family environment on children is held by the Center for Marriage and Families, based at the Institute for American Values, which claims that:

> "Family structure clearly influences educational outcomes for U.S. children. The weakening of U.S. family structure in recent decades, driven primarily by high and rising rates of unwed childbearing and divorce, has almost certainly weakened the educational prospects and achievements of U.S. children. Put more positively, there is a solid research basis for the proposition that strengthening U.S. family structure in the future – increasing the proportion of children growing up with their own, two married parents – would significantly improve the educational achievements of U.S. children." [2]

Whether or not one accepts these grim analyses, no matter whether the implied prognosis for the cognitive ability of children in the United States is on-target or terribly misguided, the point remains; virtually no one would argue that the family environment for children has substantially improved in recent years. The American public often hears from conservative pundits predicting the demise of the family, the rending of the social fabric, the loss of traditional and time-tested values, the irretrievable descent into secularity and sin, and the triumph of cultural relativism. Yet none of these pundits can explain why the measured IQ continues to rise roughly at a rate of one point every 2 years.

A recent report from the US Census Bureau makes it even harder to explain why IQ scores are rising [3]. Although the median household income remained the same from 2003 to 2004 at $44,389, the poverty rate rose from 12.5% in 2003 to 12.7% in 2004. The failure of the median household income to increase in 2004 marked the second consecutive year when there was no change in real earnings, although the cost of living continued to increase. The median earnings for men aged 15 and older who worked full-time year-round, actually declined 2.3% between 2003 and

2004, and 37.0 million people lived in poverty in 2004, up from 35.9 million people in 2003. The poverty rate for families was unchanged, but 7.9 million families lived in poverty in 2004. And the poverty rate in some parts of the country is staggering; worst among rural counties was Hidalgo, Texas, with a poverty rate of 43.6%, while worst among cities was Detroit, Michigan, with a poverty rate of 33.6%.

A great deal of research shows that poverty has a powerful effect on the home environment experienced by a child [4]. Poor families score lower on an inventory that assesses the care-giving environment for children. There is a strong relationship between poverty and what a parent can provide, though the exact nature of this relationship is not yet fully understood. When data were analyzed using a statistical approach to control poverty, race, home site, and a host of environmental and family variables, all of these things together were found to account for only about 60% of what makes each home different. This means that a large proportion of the impact of the home on a child can neither be explained nor understood. Everyone, of course, has a pet theory for that part of the home environment that has an impact on cognition, from birth order to family size to the social class of the mother. However, recent evidence suggests that birth order has little influence on a child's intelligence [5], that large family size does not necessarily produce low IQ children [6], and that parental social class explains less than 16% of the variation in a child's IQ [7].

If the family environment is getting no better for children – and arguably is getting worse – and if poverty is increasing even slightly, what would be the predicted effect upon a child's intelligence or academic performance? An extensive study – called a meta-analysis because it pooled results from 74 different component studies – found that family socioeconomic status (SES) and student academic achievement are tightly linked, with low SES predicting low achievement [8]. This finding is likely to be quite robust because conclusions are based on 101,157 students from 6,871 schools in 128 different school districts. Of all the factors that have ever been studied in the educational literature, family SES is one of the best predictors of educational success. Family SES sets the stage for student performance by directly providing resources at home and by indirectly providing the money necessary for success at school. Average family SES within a school district is the most important determinant of school financing, since roughly half of all public school funding is based on property taxes within the school district. In Illinois in 1995–1996, property tax disparity between school districts was such that the *per capita* student expenditure varied from $3,000 to $15,000. Low-SES schools typically have less to offer students than high-SES schools, in terms of teacher experience, student–teacher ratio, instructional quality, and availability of instructional materials. Family SES determines the quality of schools that a child can attend, and it can even determine the quality of relationship between parents and teachers, which can help a child to meet unusual educational needs. Overall, students at risk because of low family SES are more likely to attend schools with limited financial resources and a poor track record with students.

How can the widely-touted decline of the American family be harmonized with the firmly-established increase in IQ of American children? At present, there is no accepted rationale for why the problems that confront the American family seem to have had little or no impact on the trend for increasing student IQ. These disparate elements cannot be brought into harmony except by saying that the American family structure has a relatively small impact on measured IQ, at least at the population level. This is not to imply that SES has no impact on academic achievement, or that a strong family cannot help a child to weather personal storms. It is beyond doubt that SES affects individual achievement and that a robust family can help a child to grow. But broad social trends in the American family environment appear to have a smaller impact on IQ than does some other as-yet-unspecified factor or factors. But what might these unknown factors be?

Part of the problem may be that, through ignorance, we are constraining the notion of what constitutes environment. It is fairly obvious what the genetic influences on a child are likely to be; it is far less obvious what should be thought of as "environment."

## A New Concept of the "Environment"

Since the early days of sociology, environment has always been thought of as having to do primarily with social interactions. No less an authority than Flynn himself has written that, "it makes sense to us that the biological system determining IQ would be more stable than would be the social system determining environment" [9]. The complex and subtle interactions between nature and nurture are thus simplified to an argument about whether genes or the social network elicit a particular human trait [10]:

> "Nurture is often taken to mean the social environment that surrounds and protects the child from birth to independence. This would include early interactions with parents and siblings, as well as the more sporadic interactions with whatever members of the extended family happen to be around. Somewhat later the environment expands to include teachers and friends, and these parts of the social environment assume greater and greater importance with the passing years. Finally, those persons with whom an adolescent or young adult has lasting friendships or love relationships play an increasingly important role, whether those persons are of the same or of the opposite sex.

> But considering environmental influences to be synonymous with social influences is really very narrow and restrictive. Instead, environmental influences should be broadly defined as anything and everything not explicitly in the genes. This opens the door to many factors which might otherwise be overlooked or undervalued, including factors in the physical environment. And it also opens the door to complex interactions between genes and the environment that are neither entirely genetic nor entirely environmental. While most of these influences on behavior are still speculative, the idea that the environment has a complex and subtle impact on the individual is really not at all speculative." [10]

A great many factors that impinge on a child are clearly not genetic. In many cases, these factors are also not incorporated into the sociologist's idea of what constitutes "environment." Thus, many scientists never consider *all* of the factors that can have an impact on how a child develops or how a trait is elicited.

Where, for example, does an infection-related hearing loss fit into our tidy concept of genes *versus* environment? Hearing loss in children can result from infection with either measles virus [11] or cytomegalovirus [12]. The developing human brain is sensitive to auditory deprivation, and hearing loss in infancy can permanently impair maturation of the auditory pathway [13]. In a large group of children who were slow to learn language, roughly 13% suffered hearing loss, making impaired hearing a common cause of language delay [14]. And infection-related hearing loss can impair school performance and reduce scores on a standard IQ test [15]. Hence, despite normal genes for language ability, some children become language-impaired due to an infection. And there is now evidence that certain genes predispose a child to hearing loss after infection with cytomegalovirus [16]. Thus, genes can act to increase the individual vulnerability to variation in the environment. Conversely, variations in the environment can reveal genes that might otherwise have remained dormant. Yet relatively few sociologists have ever considered how such "non-social" parts of the environment can act to irrevocably alter the life of an individual child.

## Hypothesis: Children are Healthier and Better Able to Demonstrate Intellectual Ability

Are children simply healthier now than in the past? If children are hungry or ill or tired, they will be less able to concentrate and less able to put forth their best effort on an IQ test. If a substantial fraction of children in the "good old days" suffered from hunger or illness or some curable impairment, then those children would likely perform poorly on a test of cognitive ability, whereas children today might not be similarly impaired. Is increasing IQ perhaps a result of a rising tide of general health in the population?

## A New Concept of the Environment: The Example of Lead Pollution

A well-documented example of the environment having a large and long-lasting impact on intelligence in children is provided by lead poisoning or plumbism. Lead poisoning has been a problem for thousands of years; lead's sweet flavor was used to balance the astringency of tannin in wine, but Nikander, a Greek physician of the second century BC, recognized that colic and paralysis can follow ingestion of large amounts of lead in wine [17]. However, the realization that even very minute amounts of lead in the environment can be associated with impaired intelligence has been a long time in coming [18].

Until legislation made workplace exposure less hazardous, certain jobs were associated with exposure to very high levels of lead. Employment in battery manufacture, automobile repair, painting and paint removal, copper smelting, bridge and tunnel construction, and lead mining tended to expose workers to very high levels of lead [19]. Occupationally-exposed workers often inadvertently contaminated their own homes, thus some children were indirectly exposed to high levels of occupationally-derived lead. Nevertheless, the main cause of lead poisoning in American children was air and soil pollution from the burning of leaded gasoline, and exposure to lead-based house paint that had weathered and chipped [17].

Leaded gasoline began to be phased-out in 1976, and leaded paint was banned in 1971, but 320,000 workers in the United States were occupationally exposed to lead in 1998, and indirect exposure of children to occupational lead remains a problem. Furthermore, inner-city children – who often live in poorly-maintained homes built before the lead paint ban – can encounter high levels of lead contamination in their own home. House dust accounts for about half of a young child's total lead intake. Children are more sensitive than adults to environmental lead for many reasons: they are more likely to ingest lead particles; a child's gut absorbs lead more readily than the adult gut; and a developing brain is far more vulnerable to toxicants than a mature brain [17].

The sharp decrease in the prevalence of lead poisoning in the United States over the past 40 years is one of the greatest public health triumphs of modern times. In the 1960s, up to 20% of inner-city children had lead in their bloodstream at a level of more than 40 μg of lead per deciliter of blood (μg/dL), according to a large-scale screening of children on the East Coast [17]. From 1976 to 1980, before various regulations banning lead pollution came into effect, American children aged 1–5 years had a median blood lead level of 15 μg/dL. From 1988 to 1991, the median blood lead level fell to 3.6 μg/dL. In 1999, the median blood lead level was just 1.9 μg/dL [20]. Yet children who live in homes with lead paint can still have blood lead levels greater than 20 μg/dL, without eating paint chips.

One problem with these statistics is that blood lead levels can be a poor measure of total exposure; blood lead has a half-life in the bloodstream that is about the same as the half-life of a red blood cell – around 35 days. If a child had a high level of lead exposure 90 days ago, there might be little evidence of this exposure in the bloodstream, though the child may still suffer from impaired cognition.

A ground-breaking study in 1979 used a very ingenious method to overcome the problem of blood lead being a poor surrogate for total lead exposure [21]. In this study, parents and teachers in two towns in Massachusetts were asked to collect deciduous ("baby") teeth, as the children lost them. The idea was that lead is locked away in the bone, forming a relatively permanent marker of past lead exposure. Deciduous teeth were collected from 2,146 children, and children with the highest lead levels in their teeth were compared to children with the lowest lead levels. Children with high lead exposure had a lower full-scale IQ as well as a lower verbal IQ. They also scored poorly on tests of auditory processing and attention, and had more behavioral problems than did children with the lowest levels of lead exposure. These results were particularly striking because the study was done in the suburbs, among relatively well-to-do children, so few children had the very high levels of lead exposure that are often noted in an urban setting. This study was the first to hint that blood lead levels below 10 μg/dL could be a real problem.

There is now a great deal of evidence that blood lead levels less than 10 μg/dL predisposes a child to having a reduced IQ. A 12 year follow-up of the cohort of children who donated teeth showed that small elevations of dentine lead were associated with increased risk of school failure [22]. In addition, lead exposure was associated with reading disability, with poor class standing in high school, and with problems in fine motor control. Each 10 µg/dL increase in blood lead at 24 months of age was associated with a 5.8-point decline in full-scale IQ at school age [23]. Even at very low levels of lead exposure there were problems; full-scale IQ fell by about 7.4 points as the lifetime average blood lead increased from 1 to 10 µg/dL [24]. These results have largely been confirmed in a group of lead-exposed children in Ecuador, who were exposed to high levels of lead because their parents glazed pottery for a living [25]. Pre-industrial people may have had a level of lead exposure 100- to 1,000-fold lower than industrial-era people, since natural sources of lead exposure are rare [26]. The current understanding is that there is no threshold below which lead is non-toxic, and that lead can cause behavioral and developmental problems in addition to intellectual problems at levels far below 10 µg/dL [27].

## The Effect of Parasitic Infestation on Growth and Intelligence

Can the environment really impact children and alter human lives through a mechanism that has nothing to do with family or social interaction? Is lead an exception that does not prove the rule or can the environment profoundly impact how we think and act? Are there other non-social components of the environment that also have an impact on cognition? Do additional challenges exist in what might be called the "medical environment" that can also adversely affect a child? Is it possible that seemingly random medical events – toxicant exposures or viral infections or nutrient deficiencies – determine how a child matures cognitively?

There is another established example of the "medical environment" having a powerful impact on the cognitive ability of children; intestinal parasitic infections [28]. Between 1995 and 1996, the prevalence of intestinal parasites was measured in Sao Paulo, Brazil, by examining stool samples from over a thousand children less than 5 years of age. About 11% of the children were infected with the most common gut parasite *Giardia duodenalis*, a microorganism that infests the small intestine and spreads thorough poor sanitation. *Giardia* infection can cause symptoms of severe diarrhea, malnutrition, general malaise, and perhaps physical stunting, and it is one of the most common parasites in the world. The prevalence of *Giardia* infestation in Sao Paulo fell from 36% in 1974, to 18% in 1985, and finally to 6% in 1996, but there is still much room left for improvement. Severe infestation with *Giardia* can apparently cause physical stunting, as the growth rate of *Giardia*-infested children is significantly slower than non-infested children [29]. The effect on growth was not large – infested children grew half a centimeter less in 6 months than did non-infested children – but growth stunting was significant nonetheless. Similar studies in Turkey, where the rate of *Giardia* infection is even higher than in Brazil, found that *Giardia* infestation is associated with an eightfold increase in the risk of physical stunting and a threefold increase in the risk of slowed motor development [30].

A study of the impact of *Giardia* infestation on school performance in Turkey suggests that the parasite also reduces cognitive ability [31]. This, however, does not imply that the parasite attacks the brain; the mechanism of cognitive impairment in children with an infestation of *Giardia* has more to do with mild malnutrition and general malaise than with neurotoxicity. Yet children with *Giardia* are substantially at risk of lower school performance. A study in Peru concluded that malnutrition in infancy, which often results from infestation with *Giardia*, can result in cognitive impairment at 9 years of age [32]. This study is particularly noteworthy because it concluded that *Giardia* infestation alone is responsible for a 4-point loss in IQ on the Wechsler Intelligence Scale for Children (WISC-R), which is the same test that is often used in the United States.

Could a reduction in *Giardia* infestation account for an increase in the average population IQ? Considering the Peruvian study, which concluded that *Giardia* infestation reduced IQ by 4 points, we can project what might happen in a population. We will assume that the prevalence of *Giardia* infestation in all of Peru was the same as in Sao Paulo, Brazil [28], falling from 36% in 1974 to 6% in 1996. If we also assume that the average IQ of uninfected children is 100, then the expected population IQ in Peru would have been about 98.6 points in 1974 (i.e., 36% of Peruvian children were expected to have an IQ of 96 and 64% of children were expected to have an IQ of 100). After the prevalence of *Giardia* infestation fell to 6%, the expected population IQ in Peru would have risen to 99.8 in 1996 (6% of children were expected to have an IQ of 96 and 94% were expected to have an IQ of 100). What this means is that, by merely reducing the rate of *Giardia* infestation, Peruvian authorities potentially caused a *national rise in IQ of 1.2 points in 22 years*. This calculation, of course, ignores the fact that IQ is affected by a great many additional risk factors, and that children with *Giardia* infestation are also at risk of malnutrition. In fact, severe malnutrition is associated with a 10-point decrement in IQ, and roughly 7% of children in Peru were severely malnourished in 1999, so there could be powerful synergistic effects among the risk factors for cognitive impairment. *Giardia* infestation is thus another example of how a medical problem can cause a significant loss of IQ points in children.

## A Medical View of the Environment Through Time

We are not proposing that social environment is irrelevant; clearly, social environment still matters a great deal and it would be foolish to claim otherwise. However, we hypothesize that the medical environment is critically important in determining the growth and ultimately the intelligence of children. It may be hard to accept that improved medical care of children has caused a significant increase in the average IQ of children over the last few decades. But it may be much easier to accept that the medical environment has had a huge impact on human ecology over the past few millennia, affecting the way we live, the way we die, and even the way we think.

Primitive stone tools found in eastern Africa show that humans and human-like ancestors had the ability to cut and process meat from animals, and fossil bones often show cut marks made by stone tools more than 2 million years ago [33]. At first, humans were probably opportunistic, hunting small game, and scavenging larger game whenever possible, but later humans formed groups that hunted cooperatively and brought down larger prey. In the late Pleistocene (20,000 to 11,000 years ago), herds of animals were driven off cliffs so that hunters could obtain large quantities of meat without much risk of injury. Favored game species – such as mammoths, an ancient species of horse, and the giant gelada baboon – may have been hunted to extinction. Native American hunter-gatherers ate pronghorn antelope, bighorn sheep, and cottontail rabbit, as shown by an analysis of DNA in ancient human fecal samples [34]. In addition, fecal samples contained DNA from 4 to 8 species of plant, all consumed over a short period of time, consistent with a hunter-gatherer lifestyle.

When the shift from hunting and gathering began 10,000 years ago, there was a characteristic change in diet, with less meat, more plants, and less nutritional diversity [33]. This agricultural transition, during which humans first began to domesticate plants and animals, may have resulted from climatic change or from the extinction of a favored prey. Plants were domesticated in at least seven independent centers of civilization, and domesticated plants provided more calories per unit of land than did plants gathered wild. It has long been assumed that the transition to a more settled or "civilized" way of life was associated with a major improvement in human living conditions, but this may not have been the case. Meat is superior to plants in providing protein, calories, and certain micronutrients, and an exclusive focus on grains would have made humans vulnerable to certain forms of malnutrition. Fossil bones from around 10,000 years ago show that humans of that era were more prone to iron-deficiency anemia, with loss of bone mass, an increased rate of infection, and poor dentition, compared to human fossils of an earlier era.

As people settled in villages, there were many changes in the way they lived their lives, and these changes had an impact on patterns of disease and death. Nomadic hunter-gatherers – as our ancestors surely were – would likely have been scattered widely over the land in small groups, because this is the most efficient way to exploit resources that can be both rare and ephemeral. Small tribes or groups of people were probably less vulnerable to infectious disease, and certainly were not subject to widespread epidemics, simply because person-to-person contact outside the group was rare [35]. Nevertheless, early humans may have lived just 20 years on an average, due to the difficulty of finding food [36].

Many human diseases are thought to have arisen – or at least to have become problematic – at about the same time as people gave up the nomadic life and settled into villages [37]. Increased food availability from farming and herding would have meant that the human population could grow. People who wander in search of food live in small groups, since food sources are hard to find, whereas people who take up a stable existence may need to live in larger groups, in order to defend their resources from those who still wander. Farming and animal herding should result in a supply of food that is more reliable than hunting and gathering, and the predictability of food sources would free people somewhat from the feast-or-famine cycle of an itinerant hunter-gatherer. As farming and herding became more common, people began to live together in larger numbers and in closer proximity. Agriculture may thus have changed the ecology of human pathogens, by increasing the ease of person-to-person transmission. It has thus been postulated that disease rates increased as people became more "civilized."

Furthermore, domestication of animals may have yielded a new source of pathogens [37]. It is well known that the influenza virus is transmitted from pigs to humans, and a careful study of other human pathogens suggests that many diseases (e.g., measles, whooping cough, tuberculosis, smallpox, rabies, malaria, and intestinal parasites) evolved first in animals. Some pathogens may be unique to civilization and it is virtually certain that civilization made it easier for illnesses to move from animals to humans, and to become an epidemic. For example, since 1998 there has been a sharp increase in sleeping sickness among people in eastern Uganda, as a result of large-scale movements of the cattle that serve as a reservoir for this pathogen [38]. At least 62% of human diseases also infect animals, and roughly 77% of livestock pathogens infect more than one species, often including humans [39]. The ability to infect multiple host species is taken as an indication that a particular disease may be able to emerge as an epidemic in the alternative host, since the alternative host may have few immunological defenses against infection.

There is evidence that agriculture, which is the domestication of wild plants, can also increase the odds of disease transmission, since it often results in habitat disturbance [37]. Many new or emerging infectious diseases are associated with human modification of the environment (e.g., plague, malaria, AIDS, elephantiasis, Hanta virus, and Ebola), and it is possible that certain newly-established illnesses arose as a result of habitat disturbance.

It is interesting to note that humans domesticated a tiny fraction of the wild species that were available [40]. There are only 14 species of animal that have ever been domesticated, and only one of these domestications (the reindeer) happened in the last millennium. The five most valuable domesticated animals – the cow, pig, sheep, horse, and goat – were domesticated repeatedly, starting at least 4,000 years ago. This means that, for centuries, we have depended for our survival on a very limited selection of all available species, and this is just as true of plant domestication as it is of animal domestication. Our total dependence on a limited spectrum of food sources may account for why the agricultural transition generally meant more work for people, with smaller adult stature, poorer nutritional status, and a heavier burden of disease [40].

Study of the microscopic wear pattern of fossil teeth revealed clues of what our ancestors ate and how they lived, since microwear patterns can differentiate between diets high in plant fiber and diets high in softer, cooked foods and meat [41]. Patterns of dental pathology and tooth wear in ancient hunter-gatherers (from 9,000 BC) have been compared to the wear in more modern Neolithic (5,000–7,000 BC) people [42]. Nearly 2,000 teeth were studied for caries, ante-mortem tooth loss, dental calculus, overall tooth wear, jaw lesions, and periodontal infection. Among the hunter-gatherers, 36% had severe tooth wear and periodontal disease, but only 19% of the more modern Neolithic people had comparable problems.

Tooth wear among hunter-gatherers was extreme, especially in the most ancient people, perhaps because teeth were still used as tools and people ate highly fibrous plants. Interestingly, dental caries were rare, affecting only about 6% of hunter-gatherers and Neolithic people.

Analysis of tooth wear from the Imperial Roman era shows that tooth pathology became more common in recent times [43]. Pathological lesions (e.g., caries, abscesses, and antemortem tooth loss) as well as patterns of tooth wear were studied in 67 adults from a necropolis of the fourth century AD. There was a high frequency of caries, which likely caused the abscesses and ante-mortem tooth loss that were commonplace. There was abundant calculus and a low frequency of heavy wear in Roman teeth, which probably reflects limited consumption of fibrous foods and high consumption of carbohydrates. During later medieval times, few people were able to retain a full set of teeth past the age of 40–45 years, and many people suffered severe tooth pain, even though caries were relatively rare [44]. A change in diet in the late seventeenth and early eighteenth century measurably increased the lifespan of the dentition, but also sharply increased the prevalence of caries. This change probably also caused a major increase in tooth-related pain and suffering, and many people may have died of abscess and related causes.

It is important to note that in the "good old days" even the simplest illnesses, which today are easily cured by a course of antibiotics, could prove fatal [45]. Bones of 1,705 medieval people from the city of York, England, were examined, to determine the incidence of maxillary sinusitis. Skull bones were examined, and scientists measured the prevalence of sinusitis severe enough to cause erosion of the skull. It was found that 39% of a rural population and 55% of an urban population had severe sinusitis, which is often painful and can be fatal. The higher prevalence of sinusitis in the city may have resulted from higher levels of air pollution, due to cooking fires and perhaps industry. Studies such as these are convincing evidence that past lives were often "nasty, brutish, and short," due to the hostile effects of the medical environment.

## The Medical Environment and the Brain

We can define the "medical environment" as the sum total of all influences on individual development that are neither genetic nor social. Such a negative definition – an explanation of what medical environment is *not* – is inherently unsatisfying, but it is at least open-ended. By defining the medical environment too narrowly, we run into the risk of excluding elements that might prove to be more important than those that we know about now. Nevertheless, we know that the medical environment can include disease-causing microorganisms, persistent parasites, protein scarcities, caloric shortfalls, chronic nutrient deficiencies, environmental toxins, food-related toxicities, excessive temperatures, water shortages, altitude- or disease-related hypoxias, and environmentally-induced developmental malformations. There is even evidence of complex interactions between and among disease-causing organisms, such that having one disease can ameliorate or aggravate the symptoms of another disease [46].

The medical environment is necessarily influenced by both genetic and social factors. Some fortunate few are more resistant to pathogens or disease-causing organisms simply because of an accident of birth. For example, certain men have been identified who are able, through a poorly-understood mechanism, to resist progression of AIDS infection even after exposure to the human immunodeficiency virus [47]. A fortunate few are also resistant to pathogens or disease-causing organisms because of their social behavior. For example, some gay men are not at risk of AIDS because they altered their behavior in a way that minimized exposure to the human immunodeficiency virus.

Is the medical environment really distinct from the social environment? Clearly, both the medical and social environments are external forces that act upon an individual and can define the life course of that individual. Both the medical and the social environment are conditions over which the individual has limited control; you can avoid your father socially in the same sense that you can avoid his pathogens medically, which is to a limited degree. Both the medical and social environment interact with the individual genome, in ways that remain poorly understood, such that it is not yet possible to determine with surety to what extent a trait is a result of genes or the environment.

Yet there is a clear distinction between the medical and social environments. The social environment of an individual is usually made up of other individuals of the same species, who therefore have similar genes. To the extent that genes are shared, evolutionary interests tend to be shared as well. In most cases, the social environment is beneficial and it is rarely less than benign. In contrast, the medical environment of an individual is made up of different species, which have totally different genes. Because genes are not shared, the evolutionary interests of the organisms involved are not shared and may be completely at cross purposes. Thus, the medical environment is benign at best, and is often fatally hostile. This is a crucial distinction between the medical and social environments.

What impact do the medical and social environments have on the brain? This is a difficult question that has only a partial answer. What we know is that the brain constantly changes over the lifespan, as we learn new things and forget the old, as we mature and as we age, as we try and fail, as we grow. The brain is a highly plastic organ, specified by the genes but shaped by the environment. The purpose of the brain is to cope with the environment, to keep us alive in an often hostile world. In essence, learning is a form of behavioral flexibility, a plasticity of the brain. The brain must remain plastic, able to alter itself or be altered by the environment, for survival to be possible [48].

This perspective seems to devalue the social environment. But we would argue that the social environment has been overvalued for centuries, as a result of an incomplete notion of what can influence the individual. How a child interacts with her mother is crucial to the trajectory of life, but if the child did not survive an earlier encounter with diarrhea, parenting style is a moot issue. Tolerance and tenderness are crucial, but typhus can trump the best efforts of a parent. Educational opportunities are essential, but a child with lead poisoning or malnutrition may be unable to exploit whatever opportunities are provided. A supportive social environment may be unable to overcome the effects of a hostile medical environment.

Is this concept of the medical environment new? The medical environment has been called many things before, and sometimes it has not been called anything, but it was always present in the models built to explain behavior. Behavioral geneticists have tried to characterize the degree to which a trait is heritable or the result of shared genes, and mathematical models have been used to calculate the relative effect of genes and environment. But the models tend to measure environment without ever specifying what exactly this means, balancing genes against "shared" and "non-shared" environments. It has been assumed that the shared environment is the social interactions of the home, whereas the non-shared environment is something else, perhaps the social interactions outside the home, with teachers or with friends. But this would seem to place too much emphasis on social relationships that are often transitory or trivial. Perhaps non-shared environment is really more a measure of the medical environment than it is of transitory social encounters. This broadens the definition of environment, to incorporate all that an individual might experience, rather than just the social features of the environment.

Does this conception of the medical environment change our thinking in any way? If some feature of the environment makes people ill –be it air pollution or lead paint flaking off walls – a rational response is to change the environment. Realizing that the medical environment has an impact on cognition should encourage us to intervene, to increase the IQ of those who are impaired.

Some may argue that intervening to raise IQ is likely to be unproductive, that past efforts to raise IQ have often fallen short of expectations. Yet failure in treating a medical illness is never taken as a reason why that illness should not be treated. Heart transplantation is quite successful in treating heart failure now; roughly 2,000 patients a year receive a heart transplant, with 87% of these patients surviving more than a year and 50% surviving more than 10 years [49]. Yet the first heart transplant recipient died in only 18 days [50]. The most successful pioneer of the heart transplant surgeons, Dr. Norman Shumway, reported an actuarial survival rate for his first 29 patients as 49% at 6 months, 37% at 18 months, and 30% at 2 years [51]. Had this lack of long-term success been taken as a justification to stop heart transplantation, then tens of thousands of people would have died needlessly. Similarly, bone marrow transplantation is now a curative therapy for a wide range of diseases including leukemia, and 89% of childhood leukemia patients are cured 10 years after treatment [52]. Yet, of the first 100 pediatric leukemia patients treated by transplantation of bone marrow from a matched sibling donor, just 13 survived [53].

These findings force us to a conclusion that may be uncomfortable for political conservatives; if cognitive impairment can result from a chance encounter with the medical environment, then medical ethics requires us to intervene.

# Chapter 6
# Evidence of Physical Plasticity in Humans

There is overwhelming evidence that human IQ is increasing at an astonishing rate of about 1 point every 2 years. This rapid rate of IQ change cannot be explained as a trivial artifact, such as a flaw in the tests or a change in the way that the tests are used. Neither can this rapid rate of change be explained by evolution, a process that is inexorable but majestic in tempo. Rising IQ is also not a result of accelerated child development, since the rate of child development may not be changing and is certainly not changing fast enough to explain the Flynn effect. And clearly the social environment – the supportive matrix of interactions provided by family – is unlikely to be improving at a rate that could power a 1% increase in IQ every 2 years; in fact, many critics argue that the family environment has been getting worse for years. Thus, the only remaining possibility would seem to be that the "non-social environment" is changing, in ways that foster intellectual performance among children.

Clearly, a healthy child with better nutrition and more energy, a child with fewer school days lost to chronic illness, a child freed from the burden of hunger or homelessness, a child who has not been exposed to alcohol *in utero* or lead as an infant, a child immunized against the scourges of polio, influenza, hepatitis, and whooping cough, will be better able to shine on an IQ test. Yet it may be very hard for some people to accept that something as trivial as a vaccination program or a reduction in environmental lead can have an effect on human IQ. At issue, perhaps, is a sense that the brain determines who we are and that the essence of our being should not be so vulnerable to environmental insult; we would like to think that we are less susceptible to the sheer randomness of life. Perhaps, because each person is largely unaware of how the medical environment has impacted them, we are reluctant to credit that it can affect anyone else either. Or perhaps it is ingrained to consider the environment as being comprised entirely of the social network that surrounds us, so that it seems misguided to grant importance to a bacterium or a brief exposure to toxin.

Before we can give much credence to the idea that the medical environment has a significant impact on human cognition, it may be necessary to test some lesser points, to verify the steps in logic that led to this larger idea. Thus, it may be critical to demonstrate that the environment has an impact on the form of the body. If this claim were proven, it would make it easier to accept that the environment can also have an impact

R.G. Steen, *Human Intelligence and Medical Illness*, The Springer Series
on Human Exceptionality, DOI 10.1007/978-1-4419-0092-0_6,
© Springer Science+Business Media, LLC 2009

on the form of the brain. And if the environment can affect the form of the brain, it would become more reasonable to accept that the environment also has an impact on the function of the brain. Thus, the first step may be to show that physical plasticity – the capacity of the body to adapt in physical form over short periods of time – really can produce a notable change in human beings. But proving that the human body can change in form as a result of pressure from the environment is almost too easy.

## A Proof of Principle

The soft bones of the infant skull are easily induced to alter in shape, given slow and steady pressure, and the brain of an infant is as yielding as pudding. For thousands of years, people around the world have deliberately deformed the shape of the head in growing infants [1]. Cranial deformation is done by means of a slight but ceaseless pressure applied to the head from the first few days of life until the age of 2 or 3 years. Evidence of this custom has been found in every continent since the dawn of human civilization, from fossil Australian aboriginals of roughly 40,000 years ago [2] to parts of the modern world. Cranial molding was practiced in ancient Phoenicia by 4,000 BC and in Asian Georgia by 3,000 BC; it was described by Herodotus among people of the Caucasus and by Hippocrates among people of the Black Sea in the fifth century BC; and it has been a common practice recently in Afghanistan, Austria, Belgium, Borneo, Egypt, England, France, Germany, India, Indonesia, Italy, Malaysia, Pakistan, the Philippines, Romania, Russia, Sudan, Sumatra, Switzerland, and Zaire [1]. Yet people may be most familiar with cranial molding from the New World, where it was practiced in both North and South America.

The Olmecs, Aztecs, and Mayans of Mexico, as well as other Pre-Columbian peoples of the Andes, often molded the crania of newborns [1]. The artificially deformed skull of a person who lived more than 6,000 years ago was found in a cave in the Andes of Peru, and European visitors to Mexico described the practice thousands of years later. Small pieces of wood might be bound to the sides of the head with fabric, or the infant might be carried in a cradle-board that kept a steady pressure on the forehead. Wrappings were adjusted over time – as the infant grew or as the head shape changed – and the results were predictable enough that certain tribes could be distinguished merely by the shape of their head. The skull was modified in any of several ways: tabular compression, done by flattening one or more planes of the head, which results in a flattened forehead or an abnormally high and narrow skull; or an annular compression, done by wrapping the head tightly with a compressive bandage, which results in a conical cranial vault.

Cranial deformation is intriguing because it is a physical change induced for social reasons. Among certain Andean Indians, head shape helped to establish a person's social identity. The form of the head could identify members of the royal class, it could help to unify a tribe and define the territorial boundaries of that tribe, and it could emphasize social divisions within a society. Among the Oruro Indians of what is now Bolivia, high social-class Indians had tabular erect heads, the middle class had tabular oblique heads, and everyone else had round heads.

In the Muisca culture of Columbia, intentional cranial deformation was only done to children of the highest social class.

Surprisingly, there is no evidence that cranial deformation results in cognitive impairment. One cannot be certain of this, of course, because cognitive testing was not used extensively until recently. But, if cranial deformation had caused a marked change in cognition, it seems unlikely that people would have practiced it on their own children.

Cranial deformation proves that there can be plasticity of form in response to the physical environment. However, this does not prove that physical plasticity can result from features of the medical environment; cradle boards are applied by parents, so they could be seen as part of the social environment. Yet this is a false dichotomy, since anything that is not actually in the genes could be considered "environment," whether social or not. In a sense, both cradle-boards and trace lead in paint chips are features of the environment. The most important distinction between social and medical environment may be that we are conscious of the social matrix but are often completely unaware of the medical matrix.

Physical changes wrought by the environment clearly can occur within months or years, so the time scale is appropriate to the IQ changes that we seek to explain. Yet cranial deformation, as a proof-of-principle, may still seem facile, since it shows that the environment can modify the body under extreme circumstances, but it does not prove that physical plasticity can be induced by more subtle circumstances. Therefore, we will consider evidence that physical plasticity can occur in response to elements of the environment so subtle that people may have been unaware of the pressure to change.

## What is Physical Plasticity and How Do We Measure It?

We have said that gradual alteration of the phenotype – the physical appearance – can occur as a result of changes in the genotype – those genes that encode the phenotype. But a crucial insight is that a single genotype can produce multiple phenotypes. In other words, the environment can interact with the genotype to produce a range of different appearances or behaviors. Using the example of cranial deformation, virtually every infant with a deformed cranium had the genetic information necessary to make a normal-appearing cranium. Yet, because parents of that infant chose to bind the head in a particular fashion, the form of the head was altered in a way that was not specified by the genes. This environmentally induced change in form is physical plasticity.

But physical plasticity can also be defined as phenotypic change that occurs over historical time periods. In this context, a "historical" time period is any time period less than is required for evolution to act; thus, progressive changes in structure too rapid to be explained by evolution are a form of physical plasticity. Physical plasticity can occur over weeks or months, in the case of cranial deformation; it can occur over years or decades, in the case of environmental lead exposure; or it can occur over generations or centuries, in the case of some types of plasticity that we will discuss in this chapter.

If physical plasticity occurs over generations, this process is clearly too slow for people to be aware of, unless they have access to reliable physical measurements from prior generations. But what form could such physical measurements take? Virtually all historical records are flawed to some extent. Even measurements of height or weight are not to be fully trusted, since they may be in unfamiliar units or made with unreliable tools. Many of the important descriptors of earlier generations – caloric intake, nutrient availability, environmental toxins, or average intelligence – were not known at the time and cannot be accurately inferred after the fact. Some of the most crucial pieces of data – blood levels of trace contaminants or accurate disease diagnoses – could not be assessed until quite recently. And other important population characteristics – demographics, death rate, or average income – were recorded, but the recorded data cannot be accepted at face value. For example, measurements of the death rate during the Black Plague may be complete for a single parish, but this parish cannot be taken as a representative of what happened over a larger scale. Perhaps records were kept by a single person who was motivated by a sense that the high death rate was extraordinary, whereas other nearby parishes may not have had a motivated record-keeper or may not have been as devastated. Record-keeping was mostly done in an amateurish way in the past, without knowledge of how to collect data so that it could be analyzed accurately.

If virtually all historical records are incomplete, what are we left with? In many cases, the most reliable indicator of what people experienced is locked in the physical remains they leave behind. Science has made it possible for bones to tell stories more eloquently today than would have been possible even a few decades ago. But even the bones have problems. We cannot be sure that a particular set of bones is representative of a larger population. Thus, even if the bones of one individual are found to bear a particular trait, how can we know whether that trait was the norm? Fossils are so rare that they put scientists under a great deal of pressure to extract as much information as possible from those few fossils that have been found. Yet the scarcity of fossil evidence makes scientists vulnerable to what is called outlier bias; human attention is drawn to the novel, the unusual, the bright and glittering, and this can bias our understanding of what is "normal." The very fact that a fossil exists makes that individual highly unusual.

To put this in more practical terms, suppose we find a few fossils that are apparently fully adult human-like ancestors, but are less than 4 feet tall? Does this mean that there was a separate race or species of tiny hominids, or does this just mean that a few skeletons were, for whatever reason, much smaller than the norm? Exactly this problem has confronted scientists since the discovery of a few small-bodied hominids on the island of Flores, in eastern Indonesia [3]. The skeletons, which are fragments of perhaps eight individuals, are all about 12,000–18,000 years old, none are taller than a meter in height, and none has a brain larger than a chimpanzee. It was proposed that long-term geographic isolation led to dwarfing of what had been a *Homo erectus* stock, and that these skeletons represent a new species known as *Homo floresiensis*. The brain of this tiny "hobbit" fossil is especially intriguing; the Flores brain is only about one-third the size of a modern human brain, yet it has several features similar to a modern brain [4]. In particular, there are disproportionately large temporal lobes, which are associated with speech and hearing in humans, and large frontal lobes, which are associated with planning and executive function in humans. Even more strikingly, the surface of the Flores brain was deeply folded and gyrified, like the modern human brain. These findings were interpreted to mean that Flores man was a race derived from *Homo erectus*, able to make and use the rather advanced stone tools found nearby [4]. Yet other reputable scientists have argued, starting from exactly the same fossil material, that Flores man was actually a member of our own species, *Homo sapiens*, but that the fossilized individuals had a condition known medically as microcephaly [5].

Microcephaly is a marked reduction of brain volume that results from one of several genetic causes, and the condition is usually associated with severe mental retardation. If Flores man was actually a modern human with microcephaly, this would make it very hard to explain how these individuals were able to make and use the extensive tool kit found nearby [6]. In fact, more than 500 stone tool artifacts have been described from a nearby site, although one cannot be sure that Flores man was actually the maker of these tools, because the tools were not from exactly the same place.

Some scientists have argued that the small brain size in Flores man might have been a result of food scarcity affecting a modern human lineage [7]. Other scientists suggest that Flores man is not a microcephalic human nor is it related to *Homo erectus* or any other known ancestor, but that it is an entirely new lineage [8]. Still other scientists note that the incongruous association of a small-brained hominid with an advanced tool kit might best be resolved by considering these few individuals to be microcephalic individuals of the same species as more-typical *Homo sapiens*, with whom they shared a living site [9]. These other humans, who may have made and used the tools, were not preserved in the fossil record, perhaps through chance.

Never has there been a clearer demonstration of the difficulty of dealing with fossils than the problems associated with Flores man. Fortunately for our purposes, it is pointless to discuss the Flores remains in depth, because we have no idea what their IQ may have been. We will be able to deal with bones and fossil remains from much more recent epochs, so correspondingly more fossil material is available for analysis. With more material available, we are less vulnerable to outlier bias, and more able to determine what the "norm" truly was. Our goal will be to answer the question; does physical plasticity occur across the generations? The best way to address this question will be to look for what are called secular changes – progressive and gradual changes in physical remains.

## Unambiguous Evidence of Physical Plasticity

One of the most elegant and unambiguous studies to address the question of secular change in human remains was recently published in an obscure dental journal. This study ostensibly addressed the question of whether the form of the jaw is changing over time, perhaps as a result of changes in diet, but it really answered a far more interesting question [10]. The greatest strength of the study was that it used a large number of human remains, with every expectation that these remains represent a random sample of people from the past. Furthermore, because these remains were from two discrete time points, no guess work was needed to determine how old the remains were.

The first set of remains included 30 skulls from people who had died during the Black Death of 1348 in London, England. These 30 skulls were selected from a larger sample of almost 600 skeletons that were interred in a "plague pit," a mass grave for people who died at a time when the death rate from plague was too high for people to be buried individually. Between 1348 and 1349, the Black Death killed between a third and a half of all Londoners, so it seems likely that this pit contains a representative sample of the people who died at the time. The pit was found at a construction site at Spitalfields, near the Royal Mint, and it was excavated by archaeologists in the 1980s. Many of the 600 skeletons in the pit were disarticulated and fragmented, perhaps because bodies were flung haphazardly into the grave as workers hurried to clear dead from the streets. Yet researchers examined only the least-damaged skulls, so reliable measurements could be made. Each skull was X-rayed from the side, then measurements were made of the skull image, so it was relatively easy to standardize measurements. About 20 measurements were made of each skull, including three measurements of the height of the cranium.

The second set of remains included 54 skulls recovered from the wreck of *The Mary Rose,* the pride of King Henry VIII's fleet, which capsized and sank outside Portsmouth harbor on July 19th in 1545. The sinking of the *Rose* was witnessed by hundreds of people, including the King, so the date is beyond question. A total of 179 skeletons have been recovered, but only the most complete skulls were evaluated. One potential problem with the *Rose* sample is that it contained only one woman, whereas the plague pit was 57% female, yet this difference can be statistically adjusted. Because these skulls had been buried in anaerobic muck at the bottom of the harbor, the condition of the bones was pristine, although some of the skeletons had been disarticulated, presumably by predation.

The final sample analyzed in this study was a collection of 31 modern skull X-rays, which a university had compiled as being representative of modern people. Thus, a total of 115 skulls were evaluated in this study, rather than the one or two that are commonly reported in archaeological studies. The earliest sample was from about 33 generations ago (if we assume a generation to be 20 years), while the middle sample was from 23 generations ago. This span of time is almost certainly not long enough for evolution to have had much impact, but it is certainly time enough for secular change to occur. Evidence that the shape of the jaw differed between historical and modern samples was not overwhelming, though it seems that our forefathers may have had a more prominent mid-face than is the case today.

The largest and most convincing difference between the historical and modern samples was in the size of the cranial vault. Measuring from a landmark behind the eyes to the outer curve of the skull above the forehead, modern skulls were about 10 mm larger than the ancient skulls. This difference was highly significant in a statistical sense, as researchers calculated that the odds of this difference being random were less than 1 in 1,000. Other similar measurements made of the cranial vault were also larger in modern skulls, whereas there were no significant differences between skulls from 1348 and skulls from 1545. Thus, the cranial vault has increased in linear dimension by about 11% in modern skulls [10], which would translate into far more than an 11% increase in brain volume. There was an increase in brain volume in specifically that part of the brain (i.e., the frontal lobes) that are most concerned with reason and intellect – and this change occurred in just 459 years. A skeptic could perhaps attribute this difference to some form of cranial deformation; maybe there was a difference in the way that infants were swaddled between the Medieval Age and the present. However, nothing in the historical record suggests that this is a viable explanation. Instead, we are left with the conclusion that brain volume increased by more than 11% in less than 500 years.

There are additional studies that offer at least partial confirmation that secular change – what we call physical plasticity – is significant in other bones from other archaeological sites. Tooth size and dental arch space were studied in 48 skulls from the fourteenth to the nineteenth century in Norway, and these skulls were compared to 192 modern skulls [11]. The permanent teeth of the historical sample were smaller than those of modern people, which could account for why modern teeth are so often crowded in the jaw. Another study used X-rays to compare 31 medieval skulls from the Black Death in London to 32 modern skulls, and found that modern skulls have longer faces, longer palates, and more evidence of jaw overbite [12]. Still another study compared X-rays of 22 skulls from the Bronze Age (prior to the eight century BC) with 140 skulls from soldiers who had served in the Hapsburg Army in the late nineteenth century and with 154 contemporary recruits to the Austrian Federal Army [13]. This study found no differences between Bronze Age skulls and nineteenth century skulls (perhaps because very few ancient skulls were examined), but secular increases in skull size from the nineteenth to the twentieth centuries were clear-cut.

## Recent Evidence of Physical Plasticity

Several studies of secular change have contrasted medieval (or older) skeletons with modern skeletons, such that the span of time is quite long and even very gradual changes could be detected. Yet most of the change in skeletons seems to have been concentrated in the recent past, meaning that it might be possible to detect significant secular change even if we limit attention to recent skeletons. This is an exciting possibility, since modern skeletons are far easier to obtain.

Skulls from a forensic database – such as might be used by crime scene investigators – were combined with a set of skulls from an earlier era, to yield a total of 1,065 skulls spanning the time period from 1840 to 1980 [14]. Half the skulls in the forensic database were of people who had died by murder or suicide, so it is possible that these skulls are not a fair sample of the rest of the population. Both murder and suicide are more likely to affect those in poverty, which may explain why 41% of the skulls in the sample were from African-Americans. Nevertheless, it has never been proven that the skeletons of murder victims differ in any systematic way from the skeletons of other people, and the enormous number of skulls available for study makes this a unique data set.

The bones in this study showed clear and remarkable evidence of a progressive change in the size and shape of the cranial vault [14]. In general, the cranial vault became higher over time, as in the *Mary Rose* study. The cranium also became longer and narrower, though the shape of the face did not change much. The cranial volume increased by about 150 ml in males, which is somewhat more than a 10% increase in just 135 years. Such changes in cranial vault volume must occur during childhood, since the bones of the cranium knit together and cease growing by adolescence; thus, the vault volume is a reflection of what happened to people during their childhood. It is also noteworthy that secular change in cranial vault volume was greater than the change in body height over the same period, with height calculated from the length of the long bones. This means that the increase in brain volume cannot be dismissed as simply reflecting an increase in overall body size.

In studying physical plasticity, it is not necessary to limit our attention to skeletons alone, since X-rays or even external measurements of living people can potentially be analyzed. The father of modern anthropology, Franz Boas, who was keenly interested in the inheritance of human traits, hypothesized a century ago that the form of humans could be modified by environmental changes. To test this hypothesis, Boas gathered anthropometric data from recent immigrants to the United States, and compared immigrants to their own children. From 1909 to 1910, Boas gathered data from over 13,000 European-born immigrants and their American-born offspring in New York [15]. Many different ethnic groups are included in his sample, including Bohemians, Central Italians, Hebrews, Hungarians, Poles, Scots, and Sicilians. For each of the groups, Boas measured the head length, head width, facial width, facial height, eye color, and hair color.

Boas reported that American-born children of immigrants differ from their foreign-born parents in consistent ways [16]. Changes began in early childhood, and a crucial factor was the length of time that elapsed between when a mother first arrived in the United States and when she gave birth. Head shape tended to converge among all American-born children to a form common to Americans, especially if the mother had lived many years in the United States before giving birth. Why this is so is not known, but the numbers are quite convincing. These findings stimulated scientists to strive to understand human growth and how it is influenced by the environment, but they also stimulated a vitriolic argument about whether it is acceptable for so many immigrants to enter the country. This was the start of a controversy that continues to the present day, and some scholars have vehemently attacked Boas' methods and conclusions. With the hindsight of history, Boas' methods today seem above reproach and his conclusions well-grounded in the data:

> "...our investigations, like many other previous ones, have merely demonstrated that results of great value can be obtained by anthropometical studies, and that the anthropometric method is a most important means of elucidating the early history of mankind and the effect of social and geographical environment upon man... All we have shown is that head forms may undergo certain changes in course of time, without change of descent." [16]

Recently, Boas' data [15] on head size and shape have been analyzed again and the old controversy renewed. Two different groups evaluating the same dataset managed to come up with strikingly different conclusions. Both groups digitized Boas' data and analyzed it by computer, using far more sophisticated statistical methods than were available a century ago. The first study, by Sparks and Jantz, was highly critical of Boas, and concluded that there was a high degree of genetic influence shown by the similarity of head shape between parents and offspring [17]. The Sparks study also claimed that there was no evidence of an environmental effect and that differences between the generations were trivial, compared to differences between ethnic groups. In short, cranial morphology was not as plastic as Boas had supposed. Sparks concluded that their analysis, "supports what morphologists... have known for a long time: most of the variation is genetic variation." [17]

However, this study was flawed by a naive use of statistical tests and by a fundamental misunderstanding of what Boas actually said. The misuse of statistics would seem arcane to a non-statistician, but it leads to some confusion; the authors show two different analyses, one of which concludes that the environment had a trivial impact on the cranial shape, while the other concluded that the birth environment has a very significant impact on the head shape. Suffice it to say that the table showing a major impact of birth environment on head shape uses a stronger and more appropriate form of analysis. But a more fundamental problem with the Sparks study is that it sets up Boas as a straw man and aggressively tries to demolish his work. In fact, Boas believed strongly that genetic variation is important; in *The Mind of Primitive Man*, Boas says, "From all we know, [head measurements] are primarily dependent upon heredity."

Another critical appraisal of Boas' data concludes that Boas largely got it right [18]; cranial form is sensitive to environmental influence. Boas' enormous dataset shows unequivocally that genes do not operate in isolation from the environment. Thus, the various studies agree in showing a strong genetic influence, but there is also a strong environmental influence on head shape. The only real controversy comes in estimating the relative importance of genes and environment. However, it does not matter much, for our purposes, if the environment determines exactly 42% or exactly 53% of the variability of a particular trait. In either case, the environment has a strong impact on head shape. The extent of genetic control may differ for different traits or different people, and the precise heritability of any given trait may be fundamentally unknowable. Yet the key point is that genes and environment inevitably interact to produce the individual [19].

Changes in skull dimension are of great inherent interest, because they seem to bear directly on secular changes in IQ, but there is also evidence of secular change from other bones than the skull. Approximate body size can be calculated from the long bone of the arm, known as the humerus. A sample of over 2,500 human humeri were cobbled together by pooling historical collections with a forensic collection and a huge collection of remains from American men who fought in the Pacific during World War II [20]. This sample covers the time period from 1800 to 1970 in the United States, with the time from 1910 to 1929 especially well-represented, because that is when men who died during World War II were likely to have been born. This set of bones, because of its enormity, showed very complex trends over time. The lower limb bones of these men and women reflect a trough in body size in the mid- to late-nineteenth century, followed by a progressive secular increase in size thereafter, which lasted through the 1960s. Male secular change was more striking than female, secular change in the legs was greater than in the arms, and distal bones changed more than proximal bones (e.g., the calf increased in length more than the thigh). These changes in bone size probably reflect changes in the nutritional status of these people when they were children. Adult stature is determined in part by the difference between nutritional intake and metabolic demand, with metabolic demand calculated to include disease and physical work.

Data on the height of men inducted into the Italian Army suggests that male height is a good proxy for determining the economic well-being of people; records show that military conscripts increased progressively in height from 1854 to 1913 [21]. The increase in height was 12 cm – or roughly 4 in. – so it was not a trivial change. This height increase was not completely uniform, as there were brief periods when average height actually decreased. If height is a balance between the intake of nutrients and the claims made upon that intake, it seems reasonable to conclude that economic growth will affect both sides of that equation. A strong economy augments income, and income can be spent to buy more or better food. Economic growth also improves sanitation, medical services, and water quality, as well as reducing the need to expend calories in work. During the time period covered by Italian Army records, there was strong growth in agricultural production, which resulted in steady economic gains, and this was reflected in the increasing height of inductees.

There is now a growing consensus that human height is a surrogate measure for prosperity, that boom-and-bust business cycles are associated with cycles of human height [22]. Of course, this is not a simple relationship, since height can be affected by disease, physical activity, stress, migration, pollution, climate, and even altitude. But nutrition has a clear influence on achieved height, and physical stunting is a reliable indicator of chronic nutritional deprivation. In the United States, the link between height and prosperity is weaker among the wealthy than among the poor, as expected, since the wealthy generally do not live close to the edge of malnutrition. In addition, the relationship between height and prosperity is weaker in males than in females, for reasons unknown. In the recent era, the strength of this relationship has diminished, which may mean that people from all strata of society in the United States are now better nourished.

## Demographic Evidence of Physical Plasticity

Analyzing skull height or humerus length is an appealing way to study secular change since we can be sure that measurements were made objectively, by modern observers concerned with accuracy and precision. However, some historical records may have the same level of accuracy and precision; if used judiciously, these records can supplement what insight is available from the bones. For example, church records of births and deaths can be an objective source of data, since this information is so important both to the Church and to the families involved. Parish records on age at death can be used to reconstruct the demographic structure of a population, especially when combined with a record of births.

Roman Catholic parish records have been used to reconstruct demographic trends in central Poland over the period from 1818 to 1903, and these trends were contrasted with a demographic reconstruction of the Early Medieval period, based on 424 skeletons from a burial ground in the same parish in Poland [23]. During the medieval period, most people were tenant farmers who tilled the land of a nobleman, whereas later residents had freehold of the land they tilled. This probably resulted in better living conditions and greater food availability in the recent era, which may explain why there was a striking increase in life expectancy. During the medieval period, life expectancy for a newborn was only 25.4 years, and the average age of an adult at death was less than 40 years old. Only about 64% of people survived long enough to reach reproductive age, because of a high mortality rate among children. People who reached reproductive age realized just 69% of their overall reproductive potential, because of premature mortality. This means, at best, 45% of all people born into the Polish population had a chance to pass their genes on to the next generation. This is an astonishing statistic, which leaves the door open to more evolutionary change than might have been expected, since there was apparently very stringent selection pressure against at least some individuals in medieval Poland.

During the period from 1818 to 1903, life expectancy for a newborn increased on an average from 25 to more than 37 years [23]. If a child survived the first mortality-plagued 5 years, that child could expect to live to an average age of 53, and the age of an adult at death was almost 60 years old. Thus, average age at death increased by 50% from the Middle Ages to the modern era. Because this increase in survival coincided with major land reforms that resulted in more and better food availability, the increase in life expectancy probably reflects improved nutrition.

Another study which showed clear and unambiguous evidence of physical plasticity studied the skeletal remains of 831 children in England [24]. These remains were from archaeological excavations of three medieval cemeteries and a cemetery from the eighteenth century. Adults had been interred as well as children but, because childhood mortality was so high, nearly half the remains found were from children. Therefore, the research focus could be kept exclusively on children, since children are a cultural "canary in a coal mine," exquisitely sensitive to social and cultural changes. The goal of this research was to test whether children from the Middle Ages were as healthy as children who lived in a bustling industrial center of the eighteenth century.

The earliest cemetery evaluated was at Raunds Furnells, a rural area, which interred remains from 850 to 1100 AD. People in this area were subsistence farmers, too often close to starvation, and they lived in close proximity to cattle, which would have made them vulnerable to zoonoses (infections or parasitic infestations shared between humans and animals). A later burial site evaluated was at St. Helen-on-the-Walls, near urban York, a burial site that was active from 950 to 1550 AD. Through much of this period, York was a prosperous city; however, the particular parish where remains were found was one of the poorest, so overcrowding and poor public sanitation were likely to be problems. A third site was at Wharram Percy, a rural site near York that served as a burial ground from 950 to 1500 AD. Because Wharram Percy was only 8 miles from York, people who lived there may have been exposed to some of the same stresses as the urban population of York, though local industry was modest. The most recent archaeological site was in Spitalfields, London, which held remains interred from 1729 to 1859 AD. This site was industrial as well as urban; though parish records indicate that the congregation was rather well to do, children were probably exposed to the urban problems of overcrowding, air pollution, and inadequate sewage disposal.

For each of the 831 skeletons, age at death was estimated from the state of eruption of teeth, just as a crime scene investigator would do for recent remains [24]. Skeletal dimensions were measured, and bones were examined for specific changes indicative of metabolic stress. For example, teeth were studied, since hypoplasia or "pitting" of the enamel reveals metabolic stress in the fetal period or early childhood. Skulls were studied for evidence of maxillary sinusitis, a chronic infection of the sinuses that can erode bone, slow body growth, and even cause death. *Cribra orbitalia* was also evaluated; this is a characteristic pitting or porosity of bone in the eye socket which is indicative of iron deficiency. Systemic infection or trauma evident in the bones was recorded, as was dental disease. Finally, the subtle skeletal distortions of rickets or scurvy, which are caused by vitamin deficiencies, were noted.

Strikingly, more than half of all the remains showed *cribra orbitalia*, which is evidence of iron-deficiency anemia [24]. Roughly, a third of all children had enamel hypoplasia, revealing maternal metabolic stress during pregnancy or during the early infancy of the child. Non-specific infection was seen in nearly 20% of medieval children, but in only 4% of children from the eighteenth century. Maxillary sinusitis severe enough to cause bony erosion was seen in more than 10% of medieval children, but in only 3% of children from the Industrial Age. Yet all was not right in the later sample; 54% of children from the recent era showed metabolic diseases such as rickets and scurvy, whereas less than 20% of children from the medieval period showed evidence of metabolic disease. However, it is worth noting that all of these children were non-survivors; a very different picture might have emerged from a study of children who survived into adulthood.

This fascinating study gives a detailed and very interesting picture of the extent of physical plasticity, of how bones can be stunted or distorted by features of the medical environment. The height of rural children was greater than their urban peers by almost 3 cm, suggesting that urban children may have had an inadequate diet [24]. The very high incidence of metabolic disease in the recent era suggests that children may have been harmed by dietary fads that were common at the time. For example, during the seventeenth and eighteenth centuries, it became fashionable for women to avoid nursing their infants with colostrum, the highly nutritious form of breast milk that flows for the first few days after birth. Instead, newborns were given purges of butter, sugar, or wine, which are less nutritious and can be contaminated with illness-inducing bacteria. The resulting gastrointestinal diseases could resolve when the child was put to breast 2 to 4 days after birth, but the mortality rate in the first few days of life was quite high. Furthermore, the age at weaning from the breast fell from about 18 months to just 7 months during the Industrial Age, and children were expected to eat adult food before they were physiologically ready to do so.

Perhaps the most interesting and unexpected finding from this study is that urban living had less impact on children than did industrialization. There were trivial differences between the bones of children from urban and rural environments in the medieval era, suggesting that poor sanitation and crowding in the city did not have a major effect. But urban and rural children from the medieval period were generally healthier than children from the Industrial Age. This was particularly evident in the prevalence of rickets and scurvy, which were at least fivefold more common in the recent era. By the time that rickets was first described in 1650 AD, it was so common in the mining districts of Britain that it was called the "English disease." Rickets was at first a disease characteristic of the wealthy, because only the wealthy could afford to wean their children to fashionable formulas and condensed milk, which are low in vitamins C and D [24]. Air pollution could conceivably have aggravated a predilection to rickets, as sunlight stimulates vitamin D synthesis in the skin, but severe air pollution may have kept children indoors. In any case, industrialization had a more damaging effect on children than did urbanization, and the effects of industrialization on the skeleton are a clear testament to the plasticity of the bones.

Evidence of demographic plasticity – which probably reflects the medical environment – has now come from other archaeological sites around the world. Skeletal remains from a medieval grave site in Japan showed that life expectancy for a newborn was just 24 years, and life expectancy for a 15-year-old child was no more than 33 years [25]. This is not significantly different from an ancient Mesolithic population in Japan, implying that the life expectancy changed little for thousands of years, then improved dramatically in recent years. In Latvia, from the seventh to the thirteenth centuries, life expectancy at birth was 20 to 22 years, and life expectancy for a 20-year-old woman was 7 years less than for a comparable man [26]. This probably reflects the mortality associated with pregnancy and childbirth. On average, women gave birth to 4 or 5 children, but only half of these children survived to reach reproductive age, because of high child mortality. According to historical demographic records, average female life span has exceeded male life span only since the latter half of the nineteenth century.

## Physical Plasticity and Human Disease

We have seen evidence that the human form has changed over time, but what effect does this have in practical terms? Even if increasing height is a good surrogate for economic prosperity, what does it mean? Is height a measure of the quality of the medical environment? Are changes in physical form linked to changes in general health? Do demographic changes reflect changes in the prevalence of human disease? If people are living longer, does this mean that they are also living better or smarter? If physical plasticity shows us that human health is improving, does this necessarily mean that human intelligence is improving? The answer to some of these questions is unknown, but we do know that the medical environment can determine physical health.

Youthful exposure to an arduous medical environment can even determine the cause of death many years later. A well-documented example of the environment having a large and long-lasting impact on health is provided by the "Dutch Hunger Winter" [27]. In 1944, towards the end of World War II, after the Allies had landed in Normandy and were pushing east towards Germany, conditions in the Nazi-occupied Netherlands deteriorated badly. Food had been in short supply anyway, because there was a Dutch rail strike in September, 1944, meant to hinder Nazi troop movements and to aid the Allied advance. The Allies were able to liberate the southern part of the Netherlands, but the failure of Allied troops to capture the Rhine bridge at Arnhem slowed their advance into the northern part of the country. In retaliation for the rail strike, Nazis embargoed food transport into Amsterdam, Rotterdam, and den Hague. By the time the embargo was eased somewhat, in November of 1944, winter had come, freezing the canals and making large-scale movement of food by barge impossible.

Food intake in the Netherlands during the war had been fairly stable at about 1,800 calories per person, but food stocks in the cities dwindled rapidly; average caloric intake in Amsterdam was 1,000 calories by November and fell to less than 600 calories in January and February of 1945 [27]. As the German Army retreated, it destroyed bridges and flooded polders to slow the Allied advance which made food transport even more difficult. The famine eased somewhat with an air drop of food by the British and with importation of flour from Sweden, but the famine did not end until May, 1945, when Amsterdam was liberated by British and Canadian troops. It is thought that famine caused the death of more than 10,000 Dutch people from October, 1944 to May, 1945.

This national tragedy is unique in being a mass famine in an industrialized nation, and it has provided an historical laboratory for the study of famine for several reasons. Food shortage was circumscribed in both time and place, with famine limited to a 7-month-period in the cities, and with relative abundance of food at other times and places. Thus, children conceived or born in the cities during famine can be compared to children born elsewhere at the same time, who did not experience famine but were otherwise comparable. Furthermore, health care was excellent and birth records were meticulously kept throughout the war. Finally, the social fabric never fragmented, so the societal dislocation that occurred in other wars and in other regions did not happen in the Netherlands.

An examination of birth records in the Netherlands has shown that maternal starvation in the last trimester of pregnancy is associated with a marked reduction in birth weight [27]. Infants whose mothers endured starvation in the third trimester were smaller in weight, shorter in length, and had a reduced head circumference. In contrast, maternal starvation during the first trimester was associated with no significant reduction in birth size, and starvation in the second trimester had an intermediate effect.

The physical effects of early famine seem to last a lifetime. Blood pressure monitoring of 721 men and women aged 58 years, all of whom were born in Amsterdam during the famine, showed that these people had more marked blood pressure elevation in response to stress than is considered normal [28]. The stress-related increase in blood pressure was greatest among people who were prenatally exposed to famine early in gestation. Such stress-reactivity is expected, over the long term, to cause heart disease, and this prediction has been verified [29]. Heart health among 385 people born during the Hunger Winter was compared to that of 590 people of the same age who were not exposed to famine *in utero*. All research subjects had a physical examination and an electrocardiogram (EKG) to evaluate heart health, and 83 cases of coronary artery disease were identified. Even using a statistical adjustment to control for age, smoking history, size at birth, and social class, people who were exposed to famine *in utero* were twice as likely to have heart disease. Onset of heart disease was 3 years earlier among those who experienced famine *in utero*. People who were exposed to famine also had a higher body mass index (BMI), a more atherogenic lipid profile, and more impaired glucose tolerance, all of which suggests that they are more at risk of premature death [30]. Though mortality from diabetes or cardiovascular disease was not increased by famine exposure in one study [31], it is likely that people in that study were simply too young to show a significant mortality difference, since they were evaluated at only 57 years of age. These findings broadly support a hypothesis that chronic disease can result from stresses in the medical environment experienced during fetal life. And careful study of children born during the Dutch Hunger Winter demonstrates that the concept of "environment" cannot be limited to the family or social environment.

## Early Life Stresses and Chronic Illness

Exposure to chronic privation *in utero* or in childhood can cause health problems that last a lifetime. A study of 25,000 men and women in the United Kingdom shows that inadequate fetal nutrition increases the risk of high blood pressure, high blood cholesterol, and abnormal glucose metabolism [32]. Low birth weight increases the risk of heart disease, stroke, hypertension, and diabetes, though the precise mechanism of this risk is unknown [33]. If slow growth during infancy is combined with rapid weight gain after 2 years of age, this exacerbates the effect of impaired fetal growth. On average, adults who have a heart attack were small at birth and thin at 2 years of age and thereafter put on weight rapidly, which suggests that insulin resistance may be a risk factor for heart disease [34]. Small body size at birth is also associated with higher all-cause mortality among adult women and with premature death among adult men [35]. This effect was statistically robust, as it was based on a study of 2,339 people followed for 350,000 person-years of life. In short, heart disease in both men and women tends to reflect poor prenatal nutrition, especially if there has been "catch-up" growth during childhood [36].

For reasons that are poorly understood, early life stresses can apparently induce changes that also affect later generations. A woman's size when she is born can predict the blood pressure of her children, a full generation later [37]. These findings have led to a hypothesis that if delivery of nutrients to a growing fetus is impaired, this causes fetal metabolism to be reprogrammed in some way [38]. The fetus may adapt to under-nutrition by reducing its metabolic rate, by altering its production of or sensitivity to hormones, or by slowing its growth, and these changes can, in some way, be passed from one generation to the next. It has even been suggested that the lower infant birth weight, typical of modern African-Americans, may be an intergenerational effect of slavery [39]. This implies that the multiple generations that passed since Emancipation in 1865 have not been enough to obliterate the awful impact of slavery on the health of people in the modern era. If true, this is a powerful testament to the physical plasticity of the human body and it suggests that the medical environment can influence gene expression across the generations.

# Chapter 7
# Evidence of Mental Plasticity in Humans

We have seen evidence of a recent and rapid progressive change in the shape of the human skull – clear and unambiguous evidence of physical plasticity. We have also seen evidence of rapid secular changes in brain volume; change that is far too rapid for evolution to account for and that is consistent with an environmental influence. Insofar as it is possible to infer cause and effect after the fact, physical plasticity is a result of what we call the "medical environment." This novel concept of the environment subsumes insufficient nutrition, vitamin or nutrient deficiencies, inadequate medical care, unsatisfactory living conditions, environmental pollution, preventable parasites, treatable illnesses, and perhaps even hard physical labor during childhood. Yet so far we have only seen evidence that lead contamination and *Giardia* infection have an impact on cognition. To make a more compelling case for the medical environment, we must investigate whether other environmental features can also have an effect on cognition.

We must carefully weigh the evidence suggesting that the medical environment affects human cognition because a great deal is at stake. Unambiguous evidence that the medical environment harms our children would demand action on our part; ethically, we cannot watch the future of any child tarnished. However, action without evidence is not advisable, since it easily results in resentment and a societal backlash. If social policy is not built upon credible science and careful reasoning, it is a flimsy edifice indeed, vulnerable to a changing economy or shifting priorities.

To prove that the environment has a significant and substantial impact on human IQ over a long span of time, we must first show that mental plasticity is sufficient to produce a short-term change in cognitive ability. To be precise, we must address the question of whether brain function can be altered by the environment within a few months or a few years, and whether that alteration in cognition can persist for longer than the duration of the environmental insult. This is because, if an IQ change induced by the environment is not permanent, it may not be very important. The criterion of permanency is rigorous but necessary; environmental features that cause a transient change in IQ are much less compelling than features that cause a permanent cognitive change.

R.G. Steen, *Human Intelligence and Medical Illness*, The Springer Series
on Human Exceptionality, DOI 10.1007/978-1-4419-0092-0_7,
© Springer Science+Business Media, LLC 2009

# A Proof of Principle: Posttraumatic Stress Disorder

Proving that brain function can change as a result of the environment turns out to be far easier than one might think. Soon after the Vietnam War – when returning soldiers began to complain of having vivid and frightening "flashbacks" and when families began to note alcohol or drug abuse in men who had previously been abstinent – the diagnosis of posttraumatic stress disorder (PTSD) became deeply controversial. Clearly, PTSD cannot explain every crime or every case of drug abuse by veterans, even though there is comfort in having a diagnosis to blame for behavior. Just as clearly, it is in the financial interest of the government to deny responsibility for returning veterans who have medical problems resulting from military service.

Nevertheless, the existence of a set of symptoms that defines PTSD has not been contentious for nearly two decades [1]. The diagnosis of PTSD requires that a person experience a traumatic event that involves actual or threatened death, and that the immediate response to that trauma must involve intense fear, helplessness, or horror. Subsequently, this event is persistently re-experienced in a way that intrudes upon and interferes with daily life, through flashbacks, vivid and distressing dreams, or overwhelming anxiety. People suffering from PTSD often go to great lengths to avoid situations that remind them of the traumatic event, and they may have trouble concentrating, sleeping, or interacting with other people. Finally, these symptoms must persist for more than a month and must cause clinically significant distress. If all these conditions are met, a person can be diagnosed with PTSD, even if they have never robbed a bank while in the grip of a persistent flashback. The latest data suggest that up to 20% of soldiers returning from the Iraq war suffer from symptoms of PTSD [2]. The diagnosis of PTSD is neither controversial among clinicians nor rare among people exposed to harsh trauma.

What is more controversial about PTSD is whether it is entirely a result of the environment, or whether some people were perhaps symptomatic ("neurotic") before they were ever exposed to combat. In other words, is it still possible to blame the victim, to claim that PTSD is, in some sense, a pre-existing condition? A fairly definitive answer to this difficult question was provided by an ingenious study that focused on identical twins, who differed in their war experiences [3]. Because these twins were identical or monozygotic (MZ), their genetic liabilities were identical; hence any difference in symptoms between the twins must be due to their differing environments. By searching Army records, a total of 103 combat veterans were found, all of whom had a MZ twin who was never exposed to combat. Among combat-exposed veterans, roughly half showed symptoms of PTSD, while the rest were free of PTSD symptoms. All of the veterans and their co-twins were subjected to a series of startling sounds, while their heart rate and other measures of stress response were recorded. Among combat veterans with PTSD, the startle response was greatly exaggerated, and hearts often raced in response to threatening sounds. PTSD-plagued veterans were anxious and hyper-vigilant, and quite different from their co-twins, who tended to be more placid, even though the twins were genetically identical. Conversely, among veterans free of PTSD, there were no significant differences from their twin, who was never exposed to combat. This implies that symptoms of PTSD are an acquired illness; PTSD is due to the environment. While we cannot rule out that PTSD is associated with familial vulnerability, we can say with certainty that PTSD sufferers are no more to blame for their condition than are people with heart disease.

Recent evidence suggests that PTSD may cause physical changes in the brain. Among Dutch police officers diagnosed with job-related PTSD, a part of the brain known as the hippocampus was about 11% smaller than in officers free of PTSD [4]. In women who had PTSD as a result of severe childhood sexual abuse, hippocampi were also smaller, compared to abused women without PTSD [5]. In children who suffered sexual abuse, PTSD symptoms were also related to a loss in volume of the hippocampi [6]. In combat veterans with PTSD, hippocampal volume is reduced more in those soldiers who have a verbal impairment [7]. We, however, note that if people are assessed after they have been diagnosed of PTSD, it is impossible to distinguish between two contradictory hypotheses. Either people with small hippocampi are at greater risk of PTSD, or else severe stress causes hippocampi to wither. Proving that PTSD and atrophy both occur in the same people does not prove which condition came first. In short, correlation does not imply causality.

The issue of whether having a small brain is a cause or a consequence of PTSD is not empty semantics or a scientific shell game. Some scientists think that low intelligence increases the risk of PTSD [8, 9], and we know that people of low IQ are generally likely to have a smaller brain [10]. This suggests that the association between small hippocampi and PTSD symptoms may be more a coincidence than a cause. Some other scientists have concluded that it is the PTSD itself – rather than a history of trauma – that is associated with low IQ [11]. Still other scientists conclude that high intelligence can help people to avoid situations in which violence is likely, which would thereby decrease their risk of developing PTSD [12]. The only way to prove that brain atrophy is a result of PTSD – rather than vice versa – is to identify a cohort of people immediately after they have been exposed to extreme stress and to follow them over time. But this has not yet been done in a way that proves whether atrophy precedes or follows PTSD symptoms.

Also, considering that the hippocampus is a very small structure, measuring it with precision is difficult [13]. To compensate for errors in measurement of a very small structure like the hippocampus, and to offset natural variation in hippocampal volume, it would be necessary to examine about 250 subjects. Yet the average size of the studies cited above is far smaller (generally fewer than 30 subjects), so brain volume differences between subjects with and without PTSD may not really be meaningful. The only way to overcome this difficulty with the precision of measurement is to include more subjects in a cross-sectional study, to follow subjects in a longitudinal study as they develop PTSD, or to examine a larger part of the brain that can be measured with greater precision. These considerations are why a recent study that measured cerebellar volume in 169 children with and without PTSD [14] is so welcome; a large brain structure was measured in a large group of children. This study found that the cerebellum is indeed reduced in size, perhaps through atrophy, in children who suffer PTSD as a result of child abuse. What is especially noteworthy is that these results are valid even when they were statistically adjusted for differences in IQ between subjects. This implies that the cerebellar volume deficits in PTSD patients are a result of brain volume loss following stress, rather than a result of stress that manifests most in people of low intelligence. While this does not prove that stress causes brain atrophy, it does go some way towards confirming the hypothesis.

Yet another difficulty is that patients who develop PTSD often have comorbid disorders like alcoholism or drug addiction. Hence, it is possible that brain volume changes are a result of substance abuse, rather than of stress. A recent study tried to compensate for this by measuring the stress and alcohol consumption among 99 combat veterans with PTSD [15]. Among alcoholic PTSD patients, hippocampi were 9% smaller than normal, but in PTSD patients free of alcoholism, hippocampi were nearly normal. The absence of hippocampal atrophy in non-alcoholic patients may mean that people with PTSD only suffer atrophy if they endure the dual insult of trauma and alcoholism. Nevertheless, in either case, these results confirm our hypothesis; the medical environment can have an impact on brain structure.

But does PTSD also affect brain function? Here, the results are fairly clear-cut, because the definition of PTSD requires that symptoms of altered cognition develop soon after trauma. Patients with PTSD have decreased verbal memory, poor information processing speed, and impaired attention, even after controlling for alcohol use, education, and depression [16]. In contrast, alcoholic patients without PTSD have only decreased verbal memory. This shows that the cognitive effects of PTSD are distinct from the cognitive effects of alcoholism.

Evaluation of 299 first-graders and their caregivers suggest that exposure to urban violence is associated with a reduction of a child's IQ, relative to parental IQ [17]. This study did not assess the symptoms of PTSD, but it effectively addressed the issue of whether violence is causative of – or merely correlative with – low IQ. Since a child's IQ is usually similar to the IQ of the parents, a relative reduction in a child's IQ after encountering violence is likely due to PTSD. This study also tried to control for the home environment, socioeconomic status, and prenatal exposure to substance abuse. Of course, it is essentially impossible to know whether a statistical correction can factor out an unwanted confounder (e.g., prenatal drug exposure), but this study was strong because it included a large number of children with varying levels of violence exposure. This fascinating study – while far from the last word on the subject – concluded that PTSD following severe violence is associated with an 8-point reduction in IQ and a 10-point reduction in reading achievement. Given that the study was conducted in children whose average age was 6 years, it also shows that violence can have an impact on cognitive ability quite rapidly.

Violence is not alone in causing acute changes in brain volume. Brain volume changes over the course of a few hours, days, or weeks have also been documented in healthy people suffering dehydration from airplane travel [18]; in young people after prolonged febrile seizure [19]; in adults recovering from an eating disorder [20]; in patients with bipolar disorder receiving lithium treatment [21]; in multiple sclerosis patients left untreated [22] or treated with steroids [23]; in obsessive-compulsive patients given paroxetine [24]; in short-stature children receiving growth hormone therapy [25]; in renal failure patients who get dialysis [26]; and in schizophrenic patients treated with haloperidol [27]. Among schizophrenic patients, acute increases in brain volume occur during exacerbation of psychosis, while acute decreases in volume are linked to symptom remission [28]. Similarly, among alcoholic men, there are acute changes in brain volume during alcohol withdrawal [29, 30]. In a study of alcoholic men imaged before and again after just 1 month of abstinence, the white matter volume was found to increase by 10% [31]. These rapid changes in brain form may be accompanied by changes in brain function, which would seem to confirm that the medical environment can have a rapid – and potentially lasting – effect on brain structure and function.

## Studying Mental Plasticity

The study of mental plasticity in children has grown rapidly in recent years, in part because the tools are becoming more sensitive and in part because scientists are becoming more attuned to the issue of developmental delay in children. In the past, few would have guessed that a tiny amount of lead in the environment can cause cognitive impairment (CI). Even if some scientists had guessed it, there would have been no way to test the hypothesis; sensitive methods to characterize brain function did not exist. Now, because of cognitive tests that are able to assess brain function, and especially because of new and exquisitely sensitive methods used to image the living brain, scientists are becoming aware of just how sensitive the growing brain is to insult. Scientists now know that a range of subtle variables can have a profound impact on the growing brain and can cause CI in children.

It should also be possible to study whether cognitive enhancement can happen, but this is rarely done. To produce cognitive enhancement, scientists would have to intervene in the lives of many children, usually at great cost and without any guarantee of success. Neither taxpayers nor parents are likely to view such cognitive intervention favorably. Yet CI can result from something never intended to have an impact on cognition; lead poisoning is an accidental by-product of progress, an unplanned and insidious result of something perceived as beneficial, namely industrialization. Alternatively, CI can result from some feature of the medical environment that is so ubiquitous that we accept it; disease and malnutrition have been part of the human condition since the dawn of time. And the idea that malnutrition has a cognitive cost is a remote consideration when one is starving to death.

Finally, to be interested in studying cognitive enhancement, one has to be convinced that it is possible. Without a conviction that intervention can benefit a child, a defeatist attitude would prevail. The standard wisdom now is that intervention is fruitless, a waste of time and money; *The Bell Curve*, a shallow but widely accepted book of a decade ago merely stated what many people already believed, when it said:

> "Taken together, the story of attempts to raise intelligence is one of high hopes, flamboyant claims, and disappointing results. For the foreseeable future, the problems of low cognitive ability are not going to be solved by outside interventions to make children smarter." [32] (p. 389)

Because of such torpid fatalism, studies of cognitive enhancement have been rare, though studies of CI are not. Most studies of CI have been medical or epidemiological in nature; they began as an effort to identify the cause of some medical problem, or to determine if there are cognitive consequences of some environmental condition already known to be associated with a medical problem. For example, the physical cost of malnutrition is often obvious; the wasted bodies and lost lives, the fatal outbreaks of an illness that should not be fatal, the swollen bellies and beseeching eyes. Yet the cognitive cost of malnutrition is not at all obvious, since these costs may be hidden until years after the state of malnutrition has been corrected.

## Malnutrition and CI

Hunger is by far the biggest contributor to child mortality, and roughly half of all children who die each year are malnourished [33]. According to the United Nations, 35–40% of children suffer from moderate malnutrition and 10% from severe malnutrition during the crucial period between the second trimester of pregnancy and age 2 [34]. Save the Children has estimated that up to 79% of children in Bangladesh live in a home that is too poor to provide them an adequate diet [35].

Malnutrition –a shortage of calories or of protein – is also a significant, substantial, and preventable cause of CI. At least 130 million children – 5% of all children in developing nations around the world – are considered to have CI resulting from severe malnutrition in the first year of life [34]. And malnutrition may be more prevalent in the United States than we care to admit, given that prenatal and newborn care is unavailable for many infants, that certain pregnant women are themselves malnourished, and that there is little in the way of a safety net for at-risk infants [36].

It is hard to objectively assess the degree of malnutrition that an infant is suffering, and even harder to determine the degree of malnutrition that a person suffered in the past. Nutritionists generally agree that anthropometric measurements (e.g., height-for-age, weight-for-age, skin-fold thickness, and so on) provide the best measure of current malnutrition [37]. But some children are small even without malnutrition, hence natural variation in body size makes it hard to identify malnourished children based on body size alone. Furthermore, if relatively few children in an area are malnourished, a study will be limited by its small sample size. Thus, it is impractical to study malnutrition, except in places where the condition is quite common. And if a child has physically recovered from an episode of malnutrition during infancy, there may be little or no observable evidence of that hardship, but there may be cognitive consequences nonetheless.

Malnourishment takes many forms. Mild malnutrition may only stunt the growth of children, whereas severe malnutrition is associated with clear-cut symptoms. Severe protein deficiency – usually called kwashiorkor – can occur when children 1–3 years of age are switched from breast milk to a diet of starchy solids [37]. Children with kwashiorkor show growth failure, muscle wasting, severe depletion of blood proteins, liver hypertrophy, behavioral apathy, and edema, the latter of which produces the most distinctive symptom; a grossly swollen belly. Severe caloric deficiency – usually called marasmus – often occurs in children under 1 year of age and accompanies early weaning. Children with marasmus show marked growth failure (the child is often less than 60% of normal weight-for-age), with behavioral irritability, muscle wasting, loss of subcutaneous fat, and an emaciated appearance, but without the swollen belly of kwashiorkor. Both kwashiorkor and marasmus can co-occur with severe diarrhea, parasitic infection, or anemia, all of which further aggravate the wasting effects of malnourishment.

Children with early malnutrition typically develop CI and learning disability years later [38]. A cohort of 129 children in Barbados was identified soon after suffering moderate-to-severe protein-energy malnutrition, and these children have been followed for more than 30 years now. In fact, nearly every child born between 1967 and 1972 who suffered even a single episode of malnutrition was included in this study. When children were aged 5–11 years, their school performance was compared to that of healthy children, and deficits were identified in language, mathematics, physical sciences, social sciences, reading, religion, and crafts. Poor performance was largely accounted for by classroom inattention and, to a lesser extent, by a decrease in full-scale IQ. Follow-up of this cohort, when children were aged 11–18 years, showed that both kwashiorkor and marasmus reduce the academic performance to about the same degree [39], though other studies have concluded that marasmus, which affects younger children, is more damaging [37].

At age 11, all children in Barbados take a national high school entrance examination that is used to assign each child to either an academic or a vocational track in school [40]. Because virtually every Barbadan child takes the same examination, this test provides a well-standardized benchmark. Children with malnutrition during the first year of life scored significantly worse on this "11-plus examination." Low scores correlated with the teacher's report of classroom inattention, documented when children were as young as 5–8 years of age. Infantile stunting at 3 and 6 months also predicted poor performance on the 11-plus exam [41]. Children with stunting tend to catch up in size in later years, but CI from severe malnutrition may be irreversible [37]. Malnourished children show a loss of roughly 10–13 points in full-scale IQ, they have short attention span and poor memory, and they are easily distracted, less cooperative, and more restless than healthy children. Up to 60% of previously malnourished children suffer from attention deficit hyperactivity disorder (ADHD), whereas just 15% of healthy children show signs of ADHD. Cognitive deficits were still present at 18 years of age, and malnutrition was associated with school drop-out and poor job performance. Similar findings have now been reported in Jamaica [42], the Philippines [43], Chile [44], Guatemala [45], and Peru [46], where stunted children also scored 10 points lower on an IQ test than did children free of stunting.

Malnutrition in early childhood interacts with other risk factors for CI in unpredictable ways. For example, we have noted that malnourished infants can show symptoms of what is often diagnosed as ADHD, though it is not clear whether nutrition-related ADHD is the same as the more familiar form of ADHD. But other more subtle comorbidities are also possible. Among well-nourished Mayan infants, 79% were classified as "easy" babies by health care workers, with just 21% of infants classified as "difficult" babies [37]. This proportion of easy babies among well-nourished Mayans is comparable to the percentage of easy babies in the United States, Canada, and the United Kingdom. However, among malnourished Mayan babies, the proportion of easy babies was much lower; only 45% of malnourished babies had an easy temperament. About 55% of malnourished babies had a difficult temperament, exhibited by crying and fussiness. Irritability in a hungry infant probably has survival value, as it may elicit more care even from a tired and hungry mother. Masai infants with difficult temperaments were more likely to survive the 1974 sub-Saharan drought in Africa [37], and, even in the United States, middle-class infants who put on the most weight tend to be the most fussy [47].

In aggregate, these results show that malnutrition is a significant and preventable cause of CI. Interestingly, breast feeding is a remarkably effective way to enhance a child's IQ, compared to children who are weaned early [48]. Low-birth weight infants given breast milk got the same number of calories and grew at the same rate as formula-fed infants, but 18 months later the breast-fed infants scored better on tests of developmental maturity and cognitive ability. Infants who consume breast milk benefit by as much as 5 IQ points, compared to formula-fed infants [48]. The cognitive benefits of breast feeding appear to be durable, at least for low-birth weight infants, since breast-fed children have a higher IQ at age 3–4 years [49]. In another study of 14-year-old children, breast feeding was one of the best overall predictors of intelligence [50].

## Trace Nutrients and CI

Trace nutrient deficiency – especially of iron, iodine, zinc, vitamin A (retinol), or folic acid – is also associated with CI. More than 30% of pregnant women in developing countries are now thought to have iron deficiency anemia, and infants born with iron deficiency typically suffer from a delay in brain maturation [34]. Iron deficiency was a major problem in the United States until infant formulas were fortified with iron about 40 years ago. Even a decade ago, it was estimated that 700,000 toddlers and 7.8 million women in the United States were iron deficient, and that 240,000 toddlers had iron deficiency anemia, a degree of deficiency that results in the production of too few red blood cells [51].

Iron deficiency anemia in the United States potentially results in a substantial intelligence deficit [52]. The National Health and Nutrition Examination Survey – a study often known by the acronym NHANES – involved nearly 6,000 children between the ages of 6 and 16 years. Iron was measured using a blood sample, and children took a series of standardized cognitive tests. Among the 5,398 children who were evaluated, 3% were iron deficient, with the highest rate of iron deficiency being 8.7% among adolescent girls. Average mathematics scores were lower for those children with iron deficiency, irrespective of whether they actually had anemia. Children with iron deficiency anemia had an average mathematics score of 86.4, whereas children with iron deficiency but not anemia had a score of 87.4, and healthy children who were not iron deficient had a mathematics score of 93.7. Children with iron deficiency had more than twice the risk of scoring poorly in mathematics as did children with normal iron levels. If 3% of children in the United States are iron deficient, then roughly 1.2 million American children and adolescents may have up to a 6-point deficit in mathematical ability caused by an easily corrected iron deficiency. Interestingly, iron deficiency did not seem to result in any differences in reading ability or memory [52].

A longitudinal study of 185 infants in Costa Rica, enrolled at age 1–2 years and followed for 19 years, found that children with chronic iron deficiency never equaled children with normal levels of iron [53]. For children who were fairly prosperous but still iron deficient, the IQ was 101.2, while matched children who were not iron deficient had an IQ of 109.3. For children who were both poor and iron deficient, the average IQ was 93.1, and this cognitive gap widened as children got older. Thus, poverty and iron deficiency interact, so that a child with both problems has an IQ that is 16 points lower in childhood, and may increase to more than 25 points lower in adolescence. Across all socioeconomic levels, iron deficiency resulted in at least a 9-point deficit in IQ, and this gap worsened through adolescence and into adulthood [53].

Poor fetal iron status can identify a child who is at risk of reduced performance on cognitive tests at 5 years of age [54]. Ferritin – a protein that stores and transports iron – was measured in cord blood of 278 infants at birth, then children were cognitively tested 5 years later. Comparing children who were in the lowest 25th percentile of iron with children who were above average (< 25th to > 75th percentile), those children who were iron deficient at birth scored lower in every test. Iron deficiency was associated with weaker language ability, poorer fine-motor skills, and a somewhat lower IQ. Iron deficient infants were also nearly five times as likely to be uncoordinated. Poor iron status at birth is, therefore, a risk factor for diminished IQ, even in the United States, where we would like to think that such things can never happen.

Iodine deficiency is also a potent risk for CI. In fact, iodine deficiency may be the single most preventable cause of retardation, because iodized table salt is able to prevent most cases of iodine deficiency [34]. The *New York Times* concluded that putting iodine in salt is the most cost-effective public health measure in the world, since a ton of salt would require only 2 ounces of potassium iodate to safeguard children, and this amount of potassium iodate cost about $1.15 in 2006 [55]. However, failure to provide adequate dietary iodine lowers the intelligence by 10–15 IQ points. Worldwide, roughly 2 billion people – a third of the world's population – get insufficient dietary iodine, and this can reduce the IQ of an entire country, because iodine deficiency is often a regional problem. Nevertheless, public health measures can be successful; in Kazakhstan, where iodine deficiency was once rampant, only 29% of households used iodized salt in 1999 but 94% of households used iodized salt by 2006. It should be noted that iodine deficiency is so easily prevented that there are now fewer areas of endemic iodine deficiency than in the past; consequently, much of the science that relates iodine deficiency to CI is rather old. This should not be seen as a weakness of the science, but rather as a strength of the public health effort to increase dietary iodine.

Studies in China – where large numbers of people have had cognitive testing, both where iodine deficiency is common and where it is rare – confirm that iodine deficiency results in CI. Comparison of the distribution of test scores suggests that virtually everyone in an area of endemic deficiency may lose 10–15 IQ points [56]. Chronic iodine deficiency from dietary impoverishment also results in goiter, a condition in which the thyroid gland in the neck undergoes a painless – but occasionally monstrous – increase in volume. This can produce a grotesquely swollen neck that was, at one time, common in the United States. In eastern Tuscany, where the prevalence of goiter among school-age children was still 52% in 1990, children from an area of iodine deficiency scored significantly lower in cognitive tests than did children from an area free of goiter [57]. Even mild iodine deficiency in Tuscany was associated with increased reaction time on tests [58], suggesting that there may be permanent nerve injury from iodine deficiency. The American Thyroid Association recently published guidelines for iodine supplementation during pregnancy and lactation, because there is some evidence that – even in the United States and Canada – mild iodine deficiency is still a problem [59].

Folic acid deficiency during pregnancy can result in neural tube defects – a developmental abnormality of the brain associated with *spina bifida* and mental retardation [60]. Because folic acid supplementation during pregnancy can reduce the incidence of neural tube defects, Canada undertook a national program of fortifying cereal grain products with folic acid in 1998. After the program had been in place for 2 years, scientists compared the incidence of neural tube defects among 218,977 women who gave birth before 1998 to the incidence of birth defects among 117,986 women who gave birth after the folic acid fortification began. The rate of neural tube defects was reduced by half, a result that is clinically significant. Daily intake of just 400 µg of folic acid was a powerful protection against neural tube defects [60]. Whether folic acid supplements also increase the cognitive ability among children born with a normal nervous system is not known. However, we do know that highly intelligent children typically have in common a gene that is associated with folic acid metabolism [61], so folic acid bioavailability may determine IQ, even in people with an adequate supply of the nutrient.

Recently, greater emphasis has been laid on other trace nutrients (e.g., omega-3 fatty acids, flavonoids, vitamins B, C, and E, choline, calcium, selenium, copper, and zinc [62]), beyond the list of usual suspects related to CI (e.g., iron, iodine, zinc, vitamin A, and folic acid). It is not yet certain how these trace nutrients work to affect cognition, but it is increasingly clear that they do so; a diet deficient in any of these trace nutrient puts children at risk of impaired cognition.

## Diarrhea and CI

Children with chronic diarrhea are at risk of CI, perhaps because diarrhea makes an adequate diet marginal and a marginal diet insufficient. Severe childhood diarrhea – a leading cause of childhood mortality worldwide – also predicts impaired school performance [46]. It should not be imagined that severe diarrhea is a problem that afflicts only the Third World; in the United States in 1997 and 2000, diarrhea was the reason for 13% of hospital admissions among children less than 5 years of age [63]. This is well over 100,000 hospital admissions per year in a country where the water supply is virtually always pathogen-free. In other countries, where water may be drawn from stagnant pools at which livestock drink, the rate of diarrheal illness is astronomical. In Cambodia, one in five children has life-threatening diarrhea at any given time, largely because of unsafe drinking water. The most common intestinal parasite is *Giardia*, a microbe present in water contaminated by fecal material or by runoff from pasture land. *Giardia* attaches itself to the walls of the human small intestine in such numbers that absorption is hindered, resulting in malnutrition, trace nutrient deficiency, and stunting of an infected child. It has been estimated that up to 40% of children in developing nations show some degree of stunting [46].

Diarrheal disease severe enough to cause stunting also impairs cognitive ability. This was proven in Peru, where 239 children were studied from birth until they reached the age of 2 years, then they were evaluated for CI at age 9 [46]. At least one *Giardia* infection was diagnosed in 86% of these children, and some children had five infections per year during infancy. In the first year of life, 32% of the children were stunted, and two-thirds of stunted children were badly so, implying that their height was more than three standard deviations below the mean for that age. Children with more than one *Giardia* infestation per year scored 4 points lower than normal on an IQ test, and children with severe stunting past 1 year of age scored 10 points lower on an IQ test. Diarrhea from causes other than *Giardia* was not related to CI, so public health efforts could perhaps focus on *Giardia* [46].

Infant diarrhea causes disproportionate impairment of semantic or language fluency [64], for unknown reasons. Children with severe early diarrhea also tend to start school late and to perform poorly when they get to school [65]. Recent data from a study of 597 children in Brazil shows that *Giardia* can impair physical growth even if a child has no abdominal pain or diarrhea [66]; in these children, stunting may occur because of malabsorption of nutrients. Fortunately, the rate of *Giardia* infestation is declining in Brazil [67].

Children with *Giardia* infestation are at eightfold higher risk of stunting and threefold higher risk of delayed psychomotor development [68]. Gut parasites are common worldwide and can have a strongly negative impact on school performance and IQ [69]. One can easily imagine that the IQ of an entire nation might be reduced by parasites that could potentially be largely eliminated by a sophisticated public health effort.

## Very Low Birth Weight and CI

Very low birth weight (VLBW), defined as a birth weight less than 3.3 pounds (1,500 g), is a major public health problem in the United States. It should be noted that VLBW is not always a result of fetal malnutrition; other risk factors for low birth weight include premature birth, poor prenatal care, and perhaps certain genes. In any case, VLBW became a problem soon after hospitals established neonatal intensive care units in the 1960s, because this led to an increase in the survival of even very tiny infants. Overall, infant mortality rates have fallen from 100 deaths per 1,000 births (10%) in 1900, to just 7.2 deaths per 1,000 births (0.7%) in 1997 [70]. Poor lung function is the main factor limiting the survival of VLBW infants.

Children with VLBW tend to show CI relative to normal birth weight children [71]. VLBW children show a 10-point deficit in IQ, compared to healthy controls, with a 7-point deficit in verbal IQ and a 12-point deficit in performance IQ. There are also significant deficits in mathematics achievement, and in gross motor skills. Even 20 years after birth, problems persist for adults born at VLBW, and fewer VLBW young adults than normal graduate from high school or go on to college [72].

Infants with extremely low birth weight (less than 2.2 pounds or 1,000 grams) generally suffer more severe long-term consequences than children with VLBW [73]. When small birth weight infants were split into two groups (<750 g and 750–1,499 g) and compared to term infants at middle-school age, scientists found that achievement test scores correlated with birth weight. Very tiny infants had a reading comprehension score of 87.7, while larger VLBW infants had a reading score of 96.4, and term infants had a reading score of 102.4. Similar weight-related trends are seen for other cognitive tests, including vocabulary and mathematics, and for measures of attention, academic performance, and adaptive behavior [73]. Furthermore, the rate of mental retardation (IQ<70) at school age is higher in children born with VLBW; the rate of retardation was 21% in the lightest VLBW infants (<750 g), 8% in a heavier group (750–1,499 g), and just 2% among normal weight infants [74]. Extremely low birth weight infants also have less athletic ability, lower job competence, and less confidence as teenagers [75].

## Poverty and CI

The National Center for Children in Poverty estimates that 28.4 million children in the United States live in low-income families, with 12.8 million children living below the Federal poverty level [76]. Almost two-thirds (61%) of children from Hispanic or African-American families live in poverty, in contrast to about a quarter of children from European-American families. Research has shown that Hispanic or African-American children tend to achieve lower scores on tests of cognitive ability than do children of European-American families. Other research shows, however, that when factors related to poverty are taken into account – factors such as parental education and home environment – differences in test performance between children of color and other children are greatly reduced [77]. This implicates poverty itself as a cause of CI.

The effects of poverty on cognitive ability are virtually impossible to disentangle from the other risk factors that poverty can entail [36]. Lead exposure, poor nutrition, and increased exposure to disease have all been cited as potential causes of CI among impoverished children. Other factors may also be important, including poor parental education, overworked parents who are unable to provide care for children, and the lack of mental stimulation from books and toys. Yet another mechanism by which poverty might impair cognitive development is by heightened stress in parents and children. Living in unsafe neighborhoods, constant economic pressure from low-paying parental jobs, and family discord can all create stress in children. Chronic stress can lead to changes in the brain, and impoverished children commonly show evidence of PTSD [78].

The effect of family poverty on IQ was assessed in several studies of children [79]. When the effects of parental education and the home environment are statistically eliminated, low income still reduces IQ by up to 6 points. The timing and duration of poverty both have an effect, and the worst outcomes occur if children are poor continuously from early childhood. The home environment seems to be more strongly determined by maternal education than by poverty, which implies that better-educated mothers provide a better environment, despite conditions of poverty.

One of the factors that complicate the relationship between poverty and IQ is the correlation between parent and child IQ. Low-IQ parents tend to have low-IQ children, and are also more likely to live in poverty [80]. A great many studies have probed how genes and the environment can interact to reduce IQ [81]. For example, one study found that the degree to which IQ can be predicted by genes and environment varies according to income. In poor families, a child's IQ at age 7 was largely determined by the environment, with genes adding little [82]. However, in affluent families, the situation was reversed; genes accounted for most of the IQ variations, and the environment mattered little, perhaps because it was uniformly favorable. Thus, the role of genes is exaggerated in a favorable environment, but an unfavorable environment can overwhelm even the favorable genes in a poor family [82]. Nevertheless, it is worth noting that there is no way to separate environmental variation from genetic variation with any real precision, especially in an individual.

Another aspect of poverty that can impact a child's intellectual test performance is related to test-taking skills and motivation. Many years ago, the test-taking situation for low- and middle-income children was varied by allowing children to choose a reward before taking an IQ test [83]. Scores of low-income children increased by 11 points under conditions of social reward, compared to the scores of low-income children who were not offered a reward. In contrast, the scores of middle-income children did not change as a function of reward, suggesting that more affluent children tend to see the test itself as rewarding. This implies that lack of motivation can reduce the IQ scores earned by children of poverty. This could happen if children perceived that doing well on an IQ test was no guarantee of success; if poverty is unrelenting, it may seem as incurable as cancer.

Young children reared in poverty are likely to be deprived of many opportunities to learn; not only are the materials that foster learning absent, so too are the parents, who may have to work at several jobs to provide for the family [36]. Poverty reduces IQ through many adverse impacts on the growing brain; deprivation of high-quality nutrients, environmental toxins, chronic untreated illness, random violence leading to PTSD, and inadequate education may all play a role. It is also possible that economic hardship and the associated stresses in the "social environment" can take a toll; lack of parental time and inadequate cognitive stimulation during interactions with other adults could reduce the test performance – and the measured IQ – of impoverished children.

Poverty and associated risk factors are thought to prevent at least 200 million children in the world from attaining their full developmental potential [84]. Inadequate cognitive stimulation has been identified as a key poverty-related risk factor for CI, although it is likely that poverty entails a great many other concurrent risk factors.

## Childhood Neglect and CI

Childhood neglect – at least as scientists use the term – is not a matter of feeding your child later than expected for a night or two, and it is certainly not a matter of failing to have the latest stimulating gadgets in the cradle. Childhood neglect is more a matter of a child being subjected to an environment that fails to provide the minimal physical and emotional sustenance required for a child to thrive. An example of such a depauperate environment is provided by the state orphanages for Romanian infants abandoned at birth. These infants were subjected to extreme regimentation – all children were required to eat, sleep, and go to toilet at the same time each day – and they endured an impoverished sensory, cognitive, and linguistic environment due to unresponsive caregivers [85]. Chronic deprivation took a terrible physical toll; among 65 Romanian adoptees brought to the United States, only 15% were judged to be physically healthy and developmentally normal [86]. There was also a psychic toll; among 165 Romanian infants adopted in England, there was a pattern of failed emotional attachment, behavioral hyperactivity, attentional deficit, and quasi-autistic behavior, and such problems were so pervasive that they were attributed to institutional privation [87]. Fortunately, most adopted Romanian children were eventually able to catch up in physical growth, despite recurrent intestinal and respiratory infections [88]. Yet there are also cognitive effects of severe neglect, and these effects may not resolve following adoption; in one study, the strongest predictor of eventual cognitive function was the child's age at adoption.

A powerful recent study examined the impact of childhood neglect on Romanian orphans, and also examined the effect of environmental enrichment on cognition [85]. A total of 136 children who were institutionalized in Bucharest were identified and tested for developmental and cognitive ability. These children were then randomized either to receive continued care in an orphanage or to be moved to foster care in the home of a local family. Average age when infants were randomized was just 21 months, and children were reevaluated at 42 and 54 months of age. At both time intervals, the IQ of children in foster care was about 8 points higher than that of children who remained in the orphanages. Institutionalized children had an average IQ of 73.3 at age 54 months, whereas children in foster care had an average IQ of 81.0 at the same age, suggesting that intervention can increase cognitive ability.

However, the average IQ of matched children from Bucharest who were never institutionalized was 109.3, suggesting that even a brief institutionalization can exact a cognitive toll. Such CI could not be explained by birth weight or gestational age, but was correlated with the length of time spent in an orphanage. Because cognitive tests given to children emphasized language and verbal skills, this is clear and robust evidence that foster care can improve verbal IQ, relative to living in a state-run orphanage. In fact, the younger a child was, when moved into foster care, the better the cognitive outcome. There was evidence of a sensitive period in the first 2 years of life, so foster placement before age 2 was associated with higher IQ scores. Life in an orphanage was associated with the loss of 0.6 IQ points per month, or about 32 IQ points by the age of 54 months.

The hypothesis that brain function is impaired by childhood neglect is confirmed by a study of growth-stunted children in Jamaica. Children who are both physically stunted and emotionally neglected at age 2 years tend to achieve poorly on standardized tests of cognition by age 18 [89]. If such children receive nutritional supplements, there is no positive impact on IQ. However, if neglected children receive cognitive stimulation, they show an 8-point improvement in verbal IQ, vocabulary, and reading scores. This study is perhaps not as strong as the Romanian study, since the study design was more complicated, but it is striking that the degree of IQ increase as a result of intervention is similar.

## Lessons from Lead Poisoning

Lead is perhaps the best understood of all contributors to the medical environment, and it is proven that lead poisoning impairs cognition. Given that our knowledge and understanding of lead is relatively complete, compared to most other features of the medical environment, what lessons can we learn from lead? What does plumbism teach us that might help us to understand other features of the medical environment?

Perhaps the most crucial lesson to learn from lead poisoning is that CI is often preventable. Prior to the recognition of lead as a common environmental pollutant, only the most egregious cases of lead intoxication were recognized as a cause of CI. As we gradually became more aware of the toll of lead at high concentration, we developed sensitive tools to characterize the impact of lead at lower concentrations. Eventually, it became obvious that lead has deleterious effects at much lower levels than we knew at first. Now, it is thought that any lead in the environment at all is potentially toxic. Yet virtually all lead in the environment was put there by human activity. It is unfortunate that environmental remediation may prevent future CI, but will not affect CI that is already established.

Another key concept from lead poisoning is that there may be developmental periods during which the brain is more vulnerable to injury. There is fresh evidence from brain imaging that young children are more at risk of permanent brain injury following lead exposure [90]. In a cohort of adults whose lead exposure was carefully documented as they matured, those people with high levels of lead exposure in early childhood had more abnormality in the frontal gray matter. Had the same degree of lead exposure occurred in adulthood, damage might have been either less extensive or less permanent, since the brain is literally wiring itself during childhood. The concept of developmentally specific vulnerability makes it especially compelling to consider what environmental toxins an infant is likely to encounter. Experience would suggest that anything that creates a toxin within the home – like lead – is probably a very strong risk for infants.

Another concept that arises from a consideration of lead is that the medical environment should be considered in the context of those conditions under which humans evolved. Our species arose under conditions of competition and natural selection, which suggests that food may have been inadequate for some unfortunate children since humans first became "human." And disease pathogens are far older than the human lineage, as our primate forebears also surely fell sick and died. Hence, an argument can be made that these are "natural" elements of the medical environment. But lead was not in the environment before humans put it there, nor were air and water pollution much of a problem. Consequently, these are "unnatural" elements of the medical environment, and humans have no metabolic means to compensate for their presence. It may be true that the best we can hope for, in the modern world, is that the environment be benign and not directly harmful to a child. What we must strive to prevent is environmental toxicity, a deficit in a child's cognitive ability that results from some remediable feature of the medical environment.

The final concept that emerges from a consideration of lead poisoning is the idea that a disproportionate burden is imposed upon the poorest and those least able to fend for themselves. People plagued by poverty may be unable to afford their own homes or to adequately maintain homes that they do own, and so they are more likely to encounter lead in the environment from paint flaking off dilapidated walls. Furthermore, the poor are more likely to live near lead smelters or close to toxic dumps, where environmental pollutants are more abundant, simply because property values are lower in such places. Yet the fact that the poor are at greater risk from the consequences of capitalism means that "environmental racism" is a reality [81]:

"The future potential of a child may or may not be realized, but it is crucial for our society not to limit the future of a child simply because the past of that child has already been limited. As a society, we are not so rich in talent that we can afford to waste any of it."

# Chapter 8
# Evidence of Cognitive Plasticity in Humans

We have seen that environment can affect the form of the brain – shown unequivocally by secular changes in cranial vault volume. We have also seen that environment can affect the function of the brain – shown unequivocally in the case of stress-induced PTSD. But, we still have not demonstrated that the environment can induce a change in the cognitive capacity or intelligence quotient (IQ).

One of the surest indicators of cognitive capacity is the ability to use language. Children typically acquire language skills in just a few years, without learning formal grammar or advanced analytical thinking skills. Word learning is not just fact learning; speech perception is not just hearing; communication is not just building logical structures with simple building blocks; rather, language is a complex and highly adaptive means to convey intention, motive, and meaning. But young children learn language so rapidly that some scientists have argued that the human brain is innately adapted to language [1]. In fact, language arose in the human line, it seems to have evolved quite rapidly in humans, and it can potentially account for our tremendous success as a species. Language enables us to coordinate actions among people, thereby overcoming the individual weakness to forge a collective strength [2].

## The Importance of Language to Humans

Language is perhaps the single best indicator of human intelligence, a surrogate measure of problem-solving ability and a predictor of social aptitude. Yet, there is a growing body of evidence for plasticity in language use; in fact, human language skills are rather malleable, prone to influence by the medical environment. If language ability is indeed plastic – showing a capacity to either worsen as a result of damage done by the medical environment or to improve as a result of a successful public health intervention – this argues strongly that intelligence itself must be prone to remediation. If human IQ can increase as a result of a public health intervention, this could potentially explain the Flynn Effect. But, it could also open a new front in an ongoing culture war, between those who argue that an unfettered free market is all that any citizen should expect and those who advocate civic responsibility and the common weal. Ideas – and especially scientific data – relating to language plasticity, though sure to be contentious, are central to any discussion of what it means to live in a modern society.

We know, from a great deal of evidence, that scores on the vocabulary portion of the most widely used intelligence tests predict overall intelligence [3]. Vocabulary scores, in fact, explain more than half the total variation in full-scale IQ. This relationship is both statistically significant and clinically meaningful; vocabulary is a better predictor of full-scale IQ than any other single component of an intelligence test, more important than mathematical ability or abstract logic. A child with a weak vocabulary or another form of language impairment simply cannot be expected to perform well in a typical IQ test. To make matters worse, there is new evidence that impairments in language often co-occur with impairments in other areas, including even motor control [4].

Language is involved in social interactions and group affiliations, so a child with language impairment may also be socially impaired [4]. Learning the meaning of spoken words is a huge undertaking, driven largely by a desire to understand the intentions of others. Children with autism or other disorders of social interaction are unable – or perhaps unwilling – to understand the social aims of others; accordingly, autistic speech and vocabulary are impaired, relative to children who are socially adept. Yet autistic children are not impaired in learning all words. Words that require a form of reasoning called "mutual exclusivity" can be learned with facility – this is a form of logic that acknowledges that most objects have only one name. Rather, autistic children are specifically impaired in their ability to learn words that relate to social skills and insights. In other words, autism is not a disorder of reasoning, it is a disorder of social cognition, and impaired social cognition can impair language acquisition and life success as well.

## Genes and Language Ability

What argument can be made that language ability is *not* highly plastic? One finding that argues against plasticity is that language ability seems to be mediated by genes. A great deal of interest has recently surrounded a gene – called *FoxP2* – that seems to play a key role in language. Evidence that *FoxP2* is crucial to language emerged slowly; the first report merely described a large family with a speech disorder that seemed to be dominantly inherited, as if speech disorder were a disease [5]. Individuals from this family can have a developmental problem that limits their ability to learn language – either to understand others or to express themselves – though they do not have any other profound neurologic impairment. Only later did it emerge that several specific gene mutations are responsible for this disorder [6]. The gene in question, called the Forkhead box P2 or *FoxP2* gene, codes for a small protein that binds to DNA and controls the way other genes on the DNA molecule are regulated. Thus, *FoxP2* acts as a transcription factor, implying that it controls the way genetic information from other genes is used. *FoxP2* is expressed in the brain during development, and it appears to play a key role in building neural circuits that are essential for controlling oral muscles and generating speech [7]. There is an intriguing concordance between sites of early *FoxP2* expression in the brain and sites of later pathology in people with speech disorder. In short, the *FoxP2* gene is somehow involved in learning or executing sequences of precise motor movements, as would be required to generate the sounds of intelligible speech.

*FoxP2* is the first gene known to be relevant to the uniquely human ability to use language [8]. Interestingly, a similar gene is present in chimpanzees, gorillas, orangutans, rhesus monkeys, and even mice, and the human version of the gene differs from the mouse version by just three amino acids, albeit in a crucial part of the gene. Since humans and mice have been genetically distinct for many millions of years, these findings argue that *FoxP2* serves a vital function; otherwise evolution could have wrought change more quickly. Comparing the sequence of this gene across a range of species, it becomes evident that the human version of the gene has been the target of strong and recent natural selection during evolution. Most changes in the human gene occurred after our line separated from an ancestor shared in common with the chimpanzee. To put this in a better perspective, during 130 million years of evolution that separated chimps from mice, there was a single change in the *FoxP2* gene sequence. In the roughly 5–6 million years since humans separated from chimps, there have been two changes in the *FoxP2* sequence, a rate of evolution that is at least 43-fold faster than before the human line evolved. The best guess as to when these crucial gene changes began implies that human beings arose as a species at roughly the same time as did a gene that is required for spoken language.

## Language Impairment

People with a mutation of the *FoxP2* gene have deficits in articulation, sound production, comprehension, and grammar, as well as difficulties in producing specific sequences of oral and facial movement [9]. The brain of people with the *FoxP2* mutation appears to be normal by imaging methods such as magnetic resonance imaging (MRI), but such methods are rather insensitive. Microscopic examination of human brain tissue shows that people with the *FoxP2* mutation show subtle differences from normal, especially in certain parts of the brain that are known to be involved in speech. Aberrant *FoxP2* expression during development of a human embryo may explain structural and functional differences in the brain of at least some people who cannot articulate speech with precision at a later stage. Yet *FoxP2* is not a common cause of language impairment, suggesting that other genes – perhaps a great many other genes – must be involved in learning and producing human speech.

Specific language impairment (SLI) is defined as a failure to develop language normally, despite normal hearing, normal neurological tests, and normal exposure to human language [10]. Roughly 7% of children entering school have some degree of SLI, making this a very compelling problem for scientists concerned with childhood development. It is clear that SLI aggregates in some families, which argues that it is a result of specific genes that are aberrant in these families. Recently, an effort was made to identify genes associated with SLI, by doing a genome-wide scan of DNA collected from affected families. Early results indicate that there may be a locus on chromosome 13 that is responsible for at least some SLI. In contrast, recent results from a genome-wide study of 840 people in 184 different families [11] have identified two genes that are provisionally called *SLI1* and *SLI2*. Neither of these genes is located on chromosome 13, thereby calling into question the earlier work and highlighting some of the problems associated with gene linkage studies. Yet the evidence is reasonably strong that *SLI1* is actually associated with language impairment. This gene may also account for problems in basic reading skills, such as poor comprehension and spelling. Since such skills are important in virtually every IQ test, mutation of the *SLI1* gene could also be associated with a decrement in overall IQ.

Yet, another gene – known as *DYX1C1* – is thought to be associated with a form of language disability known as developmental dyslexia [12]. Children with a mutation of this gene have difficulty in learning to read and write, despite adequate intelligence, motivation, and education. Finding the *DYX1C1* gene thus tends to reinforce a fatalistic sense that genes can control and even limit language-learning. If language ability is truly determined by genes, how can we help children who happen to have the "wrong" genes?

## The Heritability of Language

Recent findings argue that language learning and oral communication are heritable, because genes are certainly heritable. But to what extent is language ability dependent upon genes and to what extent is it a function of the environment? As is often true, the evidence that bears most directly on heritability of a particular trait comes from studying twins [13]. Identical or monozygotic (MZ) twins share all their genes, whereas fraternal or dizygotic (DZ) twins share roughly half of their genes, with the environment more or less equally important in both MZ and DZ twins. Thus, the degree to which a trait is shared in common between MZ and DZ twins can give insight into the power of those genes that give rise to the trait. The expectation is that differences between MZ twins can only arise from the environment, whereas differences between DZ twins can arise from genes or environment. Mathematical models have been derived to calculate the "heritability" of a trait, or the proportion of variation in a particular trait that can be explained by the variation in genes. Actually, a better way to think of heritability is that it is the proportion of variation in a group of people that results from the action of a specific gene. Clearly, to make this calculation with precision, a great many MZ and DZ twins must be studied, and there must be a reliable measure of the expression of a trait of interest. But, because verbal skills can be measured reliably, the heritability of language is as well understood as is the heritability of any other trait.

Language impairment was studied in 579 twins, all of whom were 4 years old [14]. This large number of twins was allocated into 160 MZ twins, with each twin necessarily of the same sex as their co-twin, as well as 131 same-sex DZ twins, and 102 opposite-sex DZ twins. It is important to note the same-sex and opposite-sex split of DZ twins, because young girls typically perform better than young boys in language ability tests. Language impairment was defined rigorously as scoring below the 15th percentile in an extensive test of language ability. This strong study concluded that the heritability of language can vary; the more severely impaired a child is, the more likely that genes can be blamed for impairment. Among children able to function at a normal level of language proficiency, the heritability of language impairment was just 38%. But, among children who were severely impaired, the heritability of language was 76%, which may mean that the effect of "bad" genes can overwhelm even a supportive learning environment. These results show convincingly that language ability is heritable, with genes having a stronger influence at the lower end of the ability spectrum. But results also demonstrate that there is a substantial role for the environment in language ability; even among the severely impaired, environment is responsible for roughly 24% of the variation in language ability. These results have been confirmed by similar studies from the same group of scientists, which enrolled 6,963 twin pairs [15] and 9,026 twin pairs [16].

In another very clever study, scientists tested the ability of 198 same-sex twin pairs to learn nonsense words, as a way to probe their ability to learn a new language [17]. Because children in this study learned words that had no real meaning, their success could not be attributed to past learning, thereby making aptitude rather than achievement the focus of interest. This study also split twins into those with normal language ability and those who, according to their parents, had poor language ability. At 6 years of age, all twins were tested, with their relative ability adjusted for nonverbal IQ. In this fascinating study, twins with normal scores in a nonword repetition test had heritability near zero, meaning that the environment explained virtually all differences. But, for twins in which at least one child had poor language ability, heritability was 79%. These results are consistent with the previous study, even though a very different method was used, which argues strongly for the plausibility of both results.

The idea that the heritability of language is higher in children who inherit a specific genetic limitation was confirmed in a study of 4,274 twin pairs [18]. Children with small vocabularies were split into two groups; one group had average cognitive ability, whereas the other group had below-average cognition. The heritability of language problems was higher in children with a cognitive deficit; thus, the effect of environment became smaller as inherited cognitive deficits grew larger. This study also found that reading and spelling are heritable, with heritability ranging from 52% to 73% [19].

It seems that language learning – perhaps even reading and spelling – unfurls as the result of a genetic program that begins at birth. New evidence suggests that infants can recognize phonemes, the building blocks of language, within 2 months of birth [20]. Knowledge of how phonemes are used to make words accrues soon thereafter; children have an understanding of how to break a flowing sentence into meaningful words by just 12 months of age. Knowledge of how sounds interact with rules of grammar to make a sentence begins to develop at about the same age.

By 32 months of age, infants are nearly as good as adults are at recognizing when the rules of syntax have been broken. In the majority of children, language develops progressively and almost inevitably from infancy, suggesting that growth of language ability is biologically determined as surely as is growth of the human body. This sense of genetic determinism encourages us to think of language as an inherited skill, something impervious to the good or ill effects of the environment.

## Can the Environment Have an Impact on Language Learning?

The environment has a strongly negative impact on verbal reasoning in the case of fetal alcohol syndrome (FAS) [21]. Even moderate alcohol consumption by a pregnant woman can lead to cognitive impairment (CI) in a child, with specific impairment in language, and deficits can persist at least 10–15 years after exposure to alcohol. This was proven conclusively in a prospective study that enrolled more than 800 women, with alcohol use measured while women were still pregnant. About 5% of these women used alcohol nearly every day during the second and third trimester of their pregnancy, whereas 65% of the women did not drink at all during the same period. In general, the rate of alcohol consumption during the first trimester of pregnancy was higher, often because women did not know that they were pregnant. Each woman was interviewed twice while pregnant, then again when children were 8 and 18 months of age, and yet again when children were 3, 6, and 10 years of age. When children reached 10 years of age, each child underwent a thorough battery of tests to characterize cognitive ability. This is a very strong study design, because alcohol intake was measured prospectively, rather than relying upon women to recollect their alcohol consumption after the fact.

Ten years after mothers had first been enrolled in this study, over 400 children were tested to characterize cognitive ability as a function of maternal alcohol intake [21]. Alcohol exposure during the first and second trimesters significantly reduced full-scale IQ score; alcohol exposure during the first trimester resulted in a 3-point IQ deficit and exposure in the second trimester resulted in a 7-point IQ deficit. While this may not seem like much difference, few women drank more than a moderate amount of alcohol during pregnancy. Yet alcohol exposure in the second trimester still caused a 6-point decrement in verbal ability, a 6-point decrement in reasoning, a 6-point decrement in mathematical skills, and a 7-point decrement in short-term memory. Nothing else was shown to cause as large a loss in full-scale IQ as alcohol. This study is striking because at least 95% of the women behaved responsibly, according to existing knowledge when the study began in 1983.

Another recent study, which enrolled just 128 women but which followed the children for 15 years, confirmed that fetal alcohol exposure reduces the IQ of children [22]. Women in this study were recruited between 1980 and 1985, and all had at least two alcoholic drinks per week during pregnancy; in general, these women drank more than in the earlier study. On average, these women had 11–12 ounces of pure alcohol per week, equivalent to somewhat less than a bottle of scotch.

Not surprisingly, this amount of alcohol resulted in lower IQ in the offspring, as children had an 8-point deficit in full-scale and verbal IQ if they had the facial alterations typical of FAS. The greatest difference between children exposed to alcohol *in utero* and children not so exposed was in the processing speed of the brain, which suggests that there may actually be brain structural abnormalities in children with FAS.

Another study confirms that fetal alcohol exposure causes an IQ decrement, but this study was able to parse out some interesting relationships [23]. Older women are apparently at greater risk of having a child with FAS, even if they consume the same amount of alcohol as a younger woman. The reason for this increase in risk is not well understood, but it means that a woman who drinks chronically but has not yet borne children with FAS should know that her risk of having a child with FAS increases with age. Among older women, each additional ounce of alcohol per week was associated with a 2.9-point decrease in full-scale IQ. Because this study recruited 337 women while they were still pregnant, this finding is likely to be quite robust.

The degree of CI and the severity of verbal deficit are related to the total amount of alcohol consumed. Moderate alcohol consumption may result in as little as a 6-point IQ loss [21], while heavy consumption results in an 8-point deficit in full-scale and verbal IQ [22]. Very heavy alcohol consumption during pregnancy is associated with roughly a 10-point IQ deficit in children [24]. But how much deficit is expected in women who drink only wine with meals?

The risk of FAS from low levels of alcohol consumption was probed by carefully examining 543 Italian children in first-grade classes [25]. Each child received a physical examination and a battery of cognitive tests, with attention focused on children who were small for their age or who were having learning difficulties. Then the prevalence of FAS and a milder form of FAS known as fetal alcohol spectrum was determined. Overall, the prevalence of FAS and the milder spectrum disorder was higher in women who drank moderate amounts of wine than among women who abstained completely from alcohol. Children with FAS were more impaired than other children, with particular deficits in language comprehension, nonverbal IQ, and behavior. Again, results confirm that alcohol abuse in the second and third trimesters of pregnancy is worse than alcohol abuse in the first trimester. This study also suggests that FAS may be more common than has been thought, since even moderate wine consumption was potentially problematic.

Most early studies that documented the impact of fetal alcohol exposure on the IQ of children were retrospective, meaning that women were asked – often many years after the fact – to recollect their alcohol intake while pregnant. This approach has been criticized because memory is so often faulty and because women unfortunate enough to have a child with FAS may be motivated by guilt to exaggerate their alcohol use, relative to women whose children do not have FAS. It is also known that women typically admit to drinking more if they are asked after the fact than if they are asked while pregnant [26]. To help clarify which report is more accurate, pregnant women were asked prospectively to reveal their alcohol intake, then the same women were asked about alcohol intake 1 year later.

Not surprisingly, there was a difference in the amount of alcohol that women reported drinking, which proves that memory is malleable. But what is surprising is that the amount of alcohol that women admit to drinking while pregnant is lower, but still correlates better with deficits in their children. This suggests that earlier studies, which first raised a red flag about the link between alcohol abuse and FAS, underestimated the risk to a fetus from alcohol. Women who drink chronically may decrease their intake of alcohol while pregnant and later on forget the extent to which their pregnant intake was lower than normal. If women subsequently exaggerate how much alcohol they used while pregnant, it would seem that more alcohol is needed to produce FAS. In fact, harmful effects of alcohol have been found in women who drank just half an ounce of alcohol per day, which is roughly equivalent to a shot of bourbon.

The consequences of alcohol exposure for children may be permanent, according to a brain imaging study of 14 children with FAS [27]. A new form of MRI called diffusion-tensor imaging revealed characteristic differences between the FAS brain and the normal brain. Diffusion-tensor imaging shows that white matter in the brain of adolescents with FAS is abnormal, thereby suggesting that white matter structure is somehow disturbed. This could account for why the mental "processing speed" of children with FAS is below normal. However, it is not known if white matter changes are irreversible, nor even whether the imaging changes are clinically significant.

It is easy to feel superior to the mothers of FAS children; their misfortune is of their own doing and easily prevented. But this attitude is wrong for two reasons. First, the real victims of FAS are the children, who did nothing wrong. And second, the mothers themselves are often bearing a heavy burden of misfortune by the time they become pregnant [28]. A comprehensive profile of 80 mothers of FAS children showed that women were diverse in racial, educational, and economic backgrounds, but alike in being victims of abuse or challenged by mental illness. Women who drank while pregnant had lower IQ, lower income, flimsier social support networks, weaker religious affiliations, and they were more likely to have untreated mental illness. In short, these were not women who carelessly or willfully chose to endanger their children; these were women already overwhelmed by life.

It is worth noting that FAS is likely a problem of great historical significance. Alcoholism was probably more prevalent in earlier times than it is now, though there was not even an English word to describe the condition. In historical eras, water was often unsafe to drink, so drinking wine or beer during pregnancy was the lesser of two evils; FAS was unknown but cholera was achingly familiar. As Fernand Braudel wrote in *The Structure of Everyday Life* [29]:

"Drunkenness increased everywhere [in Europe] in the sixteenth century. Consumption [of wine] in Valladolid reached 100 liter per person per year in the middle of the century; in Venice, the Signoria was obliged to take new and severe action against public drunkenness in 1598... Consumption in Paris on the eve of the Revolution was of the order of 120 l per person per year.... Were there ever extenuating circumstances for this over-indulgence in wine? In fact wine, principally low-quality wine, had become a cheap foodstuff. Its price even fell relatively every time grain became too expensive... cheap calories every time bread was short." (pgs. 236–237)

# Chapter 9
# Impact of Medical Conditions on Human IQ in the United States

There is robust evidence that the IQ of young people has been rising progressively for many years. This rise in IQ – known as the Flynn Effect – is neither subtle nor recent. Careful analysis of the scientific literature suggests that the average IQ of young people is increasing at a rate of roughly 1 point every 2 years, and has probably been increasing for at least a century now. If IQ tests are not renormed frequently, the average person in 100 years would be a genius by our standards. How is this possible? Are human beings truly smarter now than they were in the past? Or do IQ tests fail to measure what they purport to measure? Are humans improving cognitively – either by a process of evolution or by an enhanced development – as rapidly as the test results would suggest? Or are test results the visible trace of a more subtle change in the human condition?

We have hypothesized that the Flynn Effect is due to the progressive amelioration of medical challenges that have depressed human intelligence for as long as there have been humans. We contend that human IQ test scores are substantially reduced by features of the environment that are medical in nature: infestation with parasites, infection with microbes, exposure to toxicants, experience with various forms of chronic deprivation, and so on.

The concept of a "medical environment" is fundamentally different from what most scientists have previously considered as environment, in that the medical environment is neither genes nor social environment. Of course, genes can have an impact on the medical environment – people with sickle cell trait are more resistant to malaria than are people who lack the sickle cell gene. Similarly, the social environment can have an impact on the medical environment – Jewish dietary laws make observant Jews unlikely to be infected with pork tapeworm, and therefore, they are not prone to neurocysticercosis.

Are these examples of the medical environment merely exceptions that fail to prove a rule, so rare as to be a distraction in explaining the Flynn Effect? Or does the medical environment have a potent effect, such that a malign medical environment can depress the human intelligence? Does the medical environment reduce IQ only in the absence of adequate medical care? Or is the medical environment a problem even in the United States, where first-rate medical care is available to all who can afford it?

R.G. Steen, *Human Intelligence and Medical Illness*, The Springer Series
on Human Exceptionality, DOI 10.1007/978-1-4419-0092-0_9,
© Springer Science+Business Media, LLC 2009

## What Medical Problems Can Impair Language Ability?

To address these questions, we collate research that addresses whether language impairment or decrements in verbal IQ can be a consequence of medical challenges, both in the United States and, in a later chapter, in the rest of the world. Rather than discussing each medical challenge in depth, it may be more compelling to provide a brief catalog of the parasites, infections, toxicants, accidents, and deprivations that are already known to reduce language ability to at least some degree. This is not intended as an exhaustive list, because this field of research is new and still developing rapidly.

To compile the list of medical conditions that cause language impairment, only two criteria were used: the condition must be proven to cause a deficit in verbal or full-scale IQ; and the condition must be remediable. These criteria were applied as rigorously as possible though, in a few cases, evidence linking a medical condition to verbal deficit is more qualitative than it is quantitative. We include specifically medical conditions and illnesses that affect children since, for the most part, it is children who take IQ tests. Though stroke can reduce verbal ability, we do not include it here, because it usually affects adults at the end of their life. We also exclude medical conditions for which no treatment exists (e.g., Downs syndrome, neurofibromatosis, fragile X), or conditions that are not usually survivable (e.g., overwhelming sepsis), since we seek to find an explanation for the Flynn effect. We omit several potential candidates from this list because the diseases are said to be eradicated (e.g., smallpox), or because too little is known about potential cognitive effects of the disease (e.g., chronic bronchitis, pneumonia, tuberculosis, cholera). Finally, cognitive effects of an illness are assumed to be chronic, even if the illness itself is curable.

Conditions that impair cognition in young people are tabulated, and a prevalence estimate for the United States is derived, often from a government entity or a disease-specific organization. The population of the United States was estimated at 300 million in 2006, with about 74 million Americans under the age of 18 years [1]. This cohort estimate was used to convert prevalence into an estimate of the number of young people affected by a condition. Conditions for which a credible IQ loss reference and a reliable prevalence estimate were obtained are then ranked by the number of IQ points potentially lost. IQ loss is calculated as the product of (Potential IQ loss per condition) × (Estimate of children at risk).

A summary of the top 30 medical conditions that potentially depress the IQ of young people in the United States is shown in Table 9.1. The estimated average IQ loss per condition is 9.0 points, with the largest decrement associated with cerebral palsy and birth asphyxia [2]. The aggregate number of IQ points potentially lost to a remediable medical challenge is about 389 million points. If there are 74 million young people in the United States [1], it is estimated that an average of 5.3 IQ points are at risk per person. Results for individual conditions are summarized.

**Table 9.1** "Maximum IQ points at risk" was calculated as (Children at risk)×(Estimated IQ risk per condition). This is done to give an idea of the scope of the problem, ignoring the fact that some children are affected by multiple medical conditions and that conditions may have additive or synergistic effects. These numbers are estimates only, as both the number of children at risk and the VIQ loss are approximated

| | Medical condition | Children at risk | Average IQ loss | Maximum IQ points at risk | IQ loss reference | Prevalence reference |
|---|---|---|---|---|---|---|
| 1 | Poverty | 10,400,000 | −4 | 41,700,000 | [4] | [4] |
| 2 | Very low birth weight | 5,700,000 | −7 | 39,900,000 | [6] | [5] |
| 3 | Attention deficit/hyperactivity | 2,886,000 | −10 | 28,900,000 | [8] | [7] |
| 4 | Lack of breast-feeding | 7,200,000 | −4 | 28,800,000 | [9] | [9] |
| 5 | Childhood cigarette smoking | 4,300,000 | −6 | 25,900,000 | [12] | [11] |
| 6 | Lead exposure (>30 mg/dL) | 3,700,000 | −7 | 25,900,000 | [13] | [13] |
| 7 | Asthma | 7,400,000 | −3 | 22,200,000 | [16] | [15] |
| 8 | Allergic rhinitis | 7,400,000 | −3 | 22,200,000 | [17] | [20] |
| 9 | Untreated bipolar disorder | 2,200,000 | −10 | 21,800,000 | [22] | [7] |
| 10 | Post-traumatic stress disorder | 2,330,000 | −8 | 18,600,000 | [23] | [7] |
| 11 | Childhood neglect | 2,100,000 | −8 | 17,100,000 | [26] | [25] |
| 12 | Untreated major depression | 5,700,000 | −3 | 17,100,000 | [27] | [7] |
| 13 | Obstructive sleep apnea | 1,500,000 | −10 | 14,800,000 | [29] | [28] |
| 14 | Iron deficiency anemia | 1,500,000 | −9 | 13,300,000 | [31] | [30] |
| 15 | PCB exposure | 1,180,000 | −6 | 7,100,000 | [35] | [34] |
| 16 | Diabetes | 1,000,000 | −6 | 6,200,000 | [43] | [39] |
| 17 | Childhood abuse | 770,000 | −8 | 6,200,000 | [44] | [25] |
| 18 | Air pollution | 1,000,000 | −6 | 6,100,000 | [46] | [45] |
| 19 | Generalized anxiety disorder | 1,520,000 | −4 | 6,100,000 | [47] | [7] |
| 20 | Mercury pollution | 5,900,000 | −1 | 5,900,000 | [49] | [48] |
| 21 | Cerebral palsy/birth asphyxia | 180,000 | −33 | 5,900,000 | [2] | [51] |
| 22 | Epilepsy | 170,000 | −12 | 2,000,000 | [54] | [52] |
| 23 | Congenital hearing loss | 74,000 | −27 | 2,000,000 | [57] | [55] |
| 24 | Congenital heart defects | 440,000 | −2 | 890,000 | [61] | [60] |

(continued)

Table 9.1  (continued)

| | Medical condition | Children at risk | Average IQ loss | Maximum IQ points at risk | IQ loss reference | Prevalence reference |
|---|---|---|---|---|---|---|
| 25 | Congenital hypopituitarism | 34,000 | −23 | 770,000 | [63] | [62] |
| 26 | Arsenic poisoning | 88,000 | −6 | 530,000 | [65] | [64] |
| 27 | Traumatic brain injury | 44,000 | −9 | 400,000 | [68] | [52] |
| 28 | Sickle cell disease | 30,000 | −12 | 360,000 | [71] | [60] |
| 29 | Human immunodeficiency virus | 10,000 | −15 | 150,000 | [73] | [72] |
| 30 | Fetal alcohol syndrome | 13,000 | −7 | 88,000 | [10] | [75] |
| | | | **Ave. loss = −9.0** | **389 million IQ points at risk** | | |
| | | | **IQ loss per child = 5.26** | | | |

**Assumptions made:**
U.S. population is 300 million, with 74 million under the age of 18 years
Cognitive effects of an illness are assumed to be chronic, even if the illness itself is curable
All calculations are based on US prevalence (# existing cases of a disease in the US at a given time)
Verbal IQ numbers are used when available; full scale IQ numbers are the second choice
When IQ scores are unavailable, the ratio of patient to control scores has been used to approximate IQ
When no IQ data are available but a cognitive deficit is likely, a 3 point deficit is assumed

1. *Poverty:* Over 42 million Americans (14.1% of the population) lived in poverty in 2005 [3], which suggests that 10.4 million American children may be cognitively impaired due to poverty. The privation of living in a disadvantaged neighborhood reduces the verbal ability of children by an average of 4 points [4].

2. *Very low birth weight (VLBW):* Roughly 7.7% of American infants are born with low or very low birth weight [5]. Children born with VLBW show deficits in cognitive ability compared to children with normal birth weight [6]. VLBW is associated with a 7-point deficit in verbal IQ, a 10-point deficit in full-scale IQ, and a 12-point deficit in performance IQ.

3. *Attention-deficit hyperactivity disorder (ADHD):* Prevalence of ADHD is 7.8% of all adults in the age group 18–29 [7]. If we assume the prevalence of ADHD in children is half that of adults, nearly 3 million children have ADHD. Verbal IQ in severely-affected children with ADHD was 20 points lower than healthy children [8], but we will assume that a 10-point deficit in verbal IQ is more common.

4. *Lack of breast feeding:* In a prospective birth cohort study of nearly 3,000 infants, 9.7% were never breast-fed [9]. Among normal-weight babies, breast feeding results in a 4-point increase in verbal IQ, if other factors are controlled [9]. We estimate that over 7 million infants in the United States show CI from premature weaning.

5. *Childhood cigarette smoking:* Maternal smoking during pregnancy reduces the verbal reasoning ability and full-scale IQ of offspring by less than 1 point [10], but when children themselves smoke there can be serious cognitive consequences. In the Americas, about 17.5% of children aged 13–15 years smoke cigarettes [11]; if we assume that this age-range is a third of the age-class, this means about 4 million children in the United States may be impaired. Comparing non-smoking children to children who smoke at least half a pack per day, verbal IQ is 6 points lower and verbal comprehension is 7 points lower among smoking children [12].

6. *Lead exposure:* Blood lead levels higher than 30 µg/dL affect 5% of children in the United States [13]. Even at very low levels of lead exposure, full-scale IQ declines by 7.4 points as lifetime average blood lead increases from 1 to 10 µg/dL [14]. There is a 6.9 point IQ decrement associated with an increase in blood lead from 2.4 to 30 µg/dL, which represents the difference between the 5th and 95th percentile [13].

7. *Asthma:* A clinical diagnosis of asthma can be made in about 10% of children [15], which amounts to 7.4 million children in the United States. Children with asthma serious enough to cause hospitalization have a 3-point decrement in full-scale IQ, relative to asthma patients who are never hospitalized [16]. Though the extent of deficit compared to healthy children is not known, we will assume that there is a 3-point verbal IQ deficit in asthma patients.

8. *Allergic rhinitis (AR):* Allergic rhinitis, which can be seasonal or perennial, affects 10–40% of Americans, is undiagnosed in about 30% of patients, and is treated with prescription medication in just 12% of cases [17]. AR is associated with CI, in patients who self-medicate with over-the-counter medications and

in patients who do not medicate. In patients who self-medicate, CI can result from the use of antihistamines that are sedating and cause cognitive slowing. In patients who do not medicate, CI can result from general malaise, insufficient nighttime sleep, and daytime sleepiness [18]. Furthermore, allergy sufferers experimentally exposed to ragweed pollen show reductions in working memory, psychomotor speed, reasoning, computation, and vigilance [19]. Early evidence suggests that recall of factual knowledge is impaired by about 10% in patients treated with a placebo (inactive) medication and by 17% in patients treated with a sedating non-prescription antihistamine [20]. Yet some studies report that AR is associated only with a perceived impairment in cognitive function, but not a measurable CI [21]. In the absence of definitive evidence, we assume that AR is associated with a degree of CI comparable to asthma.

9. *Untreated bipolar disorder (BD):* Prevalence of BD in the United States is estimated to be 5.9% among people in the age range of 18–29 years [7]. We assume that young people have a prevalence half that of the older cohort, meaning roughly 2 million young Americans have BD. A small study [22] identified a verbal IQ deficit of 22 points in BD patients; because these patients were at a state psychiatric hospital, they probably had more serious CI than usual. We estimate that the average childhood BD patient has a verbal IQ deficit of 10 points.

10. *Post-traumatic stress disorder (PTSD):* PTSD has a prevalence of 6.3% in the age class of 18–29 years [7]. If we assume that PTSD is half as prevalent in a younger cohort, then 2 million young people in the United States are affected. Among 59 children with PTSD after exposure to childhood violence, the average verbal IQ was 92 [23]. This deficit can be specifically in verbal memory, without a corresponding deficit in performance IQ [24].

11. *Childhood neglect:* A national representative survey in the United States recently concluded that 3% or roughly 2.1 million American children suffer from serious neglect or abandonment [25]. Children abandoned shortly after birth were randomized to receive institutional care or home foster care, and there was a striking difference in IQ at 33 months [26]. Institutionalized children had an IQ of 73, while children randomized to foster care had an IQ of 81. This suggests that childhood neglect is ameliorated by human contact.

12. *Untreated major depression:* Depression affects 15.4% of young adults [7]; if the prevalence of depression is half this high in young people, then about 6 million American young people suffer depression. Depressed patients who remit show a 3 point increase in IQ, and improved cognition may be a marker of recovery [27].

13. *Obstructive sleep apnea (OSA):* OSA has a prevalence of about 2% among children [28]. Severe chronic OSA is associated with a 15-point deficit in full-scale IQ with specific problems in verbal and working memory, relative to healthy children [29]. It is not known whether typical apnea or snoring is associated with any deficit, hence we will assume that milder sleep apnea results in a 10-point decrement in verbal IQ, perhaps because of daytime sleepiness.

14. *Iron deficiency anemia:* Iron deficiency is prevalent in 2% of young children in the United States [30]. Teenagers identified as iron deficient during infancy had a full-scale IQ 9 points lower than children at age 19, and this gap widened to 25 points if iron deficiency co-occurred with poverty [31]. Prenatal iron deficiency is also associated with a threefold higher risk of mental retardation [32].

15. *Polychlorinated biphenyl (PCB) exposure:* PCBs are organic chlorine chemicals which accumulate in the environment and in living tissue [33]. A national survey in 2002 suggested that 5.5% of Americans have elevated PCB levels and 29% of these people have undiagnosed diabetes [34]. We will assume that PCBs impair cognition only in people with undiagnosed diabetes, but this still means that roughly 1 million young Americans are at risk. Prenatal exposure to PCBs can cause a significant reduction in verbal and full-scale IQ, with IQ scores of the most highly exposed children 6 points lower than healthy children [35]. In New York, adolescents exposed to PCBs showed impairment of reading, long-term memory, and general knowledge [36]. Among Dutch children exposed to PCBs in mildly polluted water, *in utero* exposure to PCBs was associated with poor cognitive function in preschool [37]. Yet other studies have been unable to replicate these findings, so this research is highly controversial [38].

16. *Diabetes:* The prevalence of diabetes was 2.8% in 2000, and is estimated to rise to 4.4% in 2030, if obesity continues to become more prevalent [39]. Well-controlled childhood diabetes may not produce a decrement in academic performance [40], but well-controlled diabetes requires high-quality medical care. New evidence suggests that diabetes can impair glucose delivery to a part of the brain called the hippocampus, causing an impairment of memory formation [41]. In children with poorly-controlled diabetes or a history of hypoglycemic seizures, there can be reduced verbal comprehension, with impaired short-term and word memory [42]. Six years after onset of childhood diabetes, verbal IQ scores were 6 points lower than in control children, with a 5-point decrement in full-scale IQ [43].

17. *Childhood abuse:* A recent national survey concluded that 10% of American children suffer physical abuse [25]; thus 7.4 million young people are at risk. Exposure to domestic violence can reduce IQ; this conclusion is based on a study of 1,116 twins, and hence, is probably quite robust [44]. Children exposed to low levels of domestic violence show a reduction of 1 point in IQ, while children exposed to high levels of domestic violence show an 8 point reduction [44]. We assume that roughly 10% of abused children suffer injury serious enough to result in an 8 point IQ decline.

18. *Air pollution:* Air pollution takes many forms, but the main form considered here is exposure to polycyclic aromatic hydrocarbons (PAH). About 2.8% of urban infants suffer from severe exposure to PAHs [45], and we estimate that half of all infants are urban, so roughly 1 million young people may have been exposed to damaging levels of PAH. "Personal air sampling" near pregnant women can predict which infants will suffer developmental delay, and the difference in IQ between children with high- and low-exposure to PAH was 6 IQ points [46].

19. *Generalized anxiety disorder (GAD):* Prevalence of GAD in the United States is 4.1% in the age range of 18–29 years [7]. If the prevalence is half as high in younger people, then 1.5 million young Americans have GAD. Full-scale IQ of 57 children with anxiety disorder was 4 points less than in 103 healthy matched children [47].

20. *Mercury pollution:* Mercury is released by coal-fired power plants and by waste incineration, and mercury contaminates seafood. Roughly 8% of American women have blood mercury levels higher than recommended [48], suggesting that 5.9 million young Americans are at risk. At the highest levels of contamination in the United States, it is likely that no more than 3 IQ points are lost to mercury toxicity [49]. The best current evidence suggests that dietary mercury may be responsible for roughly a 1-point IQ deficit [50].

21. *Cerebral palsy (CP) and birth asphyxia:* Cerebral palsy often results from poor prenatal or perinatal medical care. Prevalence of CP in children aged 3–10 years is 2.4 per 1,000 children [51], so about 180,000 American young people have CP. In a random sample of 68 children with CP, the average full-scale IQ was just 67 [2].

22. *Epilepsy:* Lifetime prevalence of epilepsy in the United Kingdom is about 4 per 1,000 people [52]. If prevalence in the United States is similar, then 300,000 young Americans are epileptic. Not all epilepsy is preventable, but epilepsy can result from preventable causes, including meningitis, tuberculosis, or head injury. Among 1,017 children with epilepsy, there was no evidence of congenital brain damage in 57% of cases [53], so we estimate that 57% of epilepsy cases could be prevented. Epileptic children score lower than healthy children on all cognitive measures, with scores about 12 points lower on verbal IQ [54].

23. *Congenital hearing loss:* Congenital bilateral permanent hearing loss affects roughly 1 infant for every 1,000 births [55], or 74,000 young people in the United States. Hearing impairment is specifically associated with a deficit in verbal IQ [56]. Scores in language ability were 27 points lower in children with profound hearing loss, compared to children mildly impaired by hearing loss [57]. In a study of 122 deaf pupils in the Netherlands, preventable causes of deafness were more common than inherited causes, and every deaf child had moderate-to-severe retardation, which may simply reflect the difficulty of measuring IQ in a deaf child [58]. Some clinicians argue that IQ in deaf children is actually comparable to children with normal hearing, if testing is done so as not to discriminate against hearing disability [59], but life discriminates nonetheless.

24. *Congenital heart defects:* The prevalence of congenital heart defects is 4 to 8 births per 1,000 in the United States [60]. We assume that 6 children per 1,000 have congenital heart defects and survive surgical treatment, which means over 400,000 young people have treated heart defects in the United States. A study of 243 children treated with surgery for congenital heart defects found that verbal IQ was 2 points less than normal [61].

25. *Congenital hypopituitarism:* Prevalence of hypopituitarism is estimated to be 46 cases per 100,000 people [62], or 34,000 young people in the United States.

The degree of CI in congenital hypopituitarism is striking; among 10 treated children, verbal IQ was 23 points less than normal [63]. This deficit could arise through abnormal brain development or through low blood glucose or growth hormone levels in infancy.

26. *Arsenic poisoning:* Arsenicosis typically results from drinking contaminated water, and it has been estimated that about 350,000 people in the United States drink water containing more than 50 ppb [64]. If we assume that 25% of those exposed were young, this amounts to 88,000 young Americans. Childhood exposure to high levels of arsenic can reduce verbal IQ by 6 points [65], and children with high levels of arsenic are impaired in verbal ability [66].

27. *Traumatic brain injury (TBI):* An epidemiological study of severe brain injury in the United Kingdom estimated that there are 60 cases per 100,000 people [52], or roughly 44,000 young Americans. About 23% of emergency department visits in the United States result from TBI with loss of consciousness [67]. When children with TBI return to school, two-thirds have difficulties with school work, and there can be problems with memory and attention. Among 56 brain-injured children, the average verbal IQ score was 91 [68]. Among children brain-injured as infants, 48% had an IQ below the 10th percentile [69].

28. *Sickle cell disease (SCD):* Roughly, 40 in every 100,000 births in the United States have hemoglobin disorders [60], and the most common disorder is SCD. About 30,000 young people in the United States have SCD, and cognitive function of children with SCD is impaired, even among children free of stroke [70]. In 54 children with SCD, verbal IQ was 12 points lower than healthy children matched for age, race, and gender [71].

29. *Human immunodeficiency virus (HIV):* About 10,000 young people in the United States have HIV infection [72]. Children with clinically stable HIV have a verbal IQ of 82–85 [73]. Transmission of AIDS from mother to child can be prevented in at least 85% of HIV-infected mothers, if mothers are given antiviral treatment [74].

30. *Fetal alcohol syndrome (FAS):* FAS was diagnosed in 17 per 100,000 births in the United States in 2002, which represents a 74% reduction in the number of newborns with FAS over the past decade [75]. Nevertheless, there are likely to be 13,000 young people with FAS in the United States. Alcohol exposure in the second trimester results in an approximate 7-point IQ deficit [10].

## Can Cognitive Impairment from Poverty Be Prevented?

Poverty is the single most important public health challenge that reduces verbal IQ in the United States. Moreover, poverty can potentially interact with other conditions, including low birth weight, ADHD, lack of breast feeding, cigarette smoking, lead exposure, asthma, AR, PTSD, childhood neglect, depression, and so on [76], to further reduce childhood IQ.

A thorough longitudinal study of over 2,000 children living in disadvantaged neighborhoods in Chicago proved that poverty has a chronic effect on the verbal ability of African-American children [4]. Children were tracked for up to 7 years as they moved into or out of a poverty-stricken area, to correct for changes in the environment that each child might experience as they move. Sustained poverty reduced verbal ability by more than 4 points, roughly equivalent to missing an entire year of school [4].

It is hard to know how much poverty-related cognitive impairment (CI) is preventable. However, a plausible estimate can be obtained by considering how much CI can be remediated by an aggressive educational intervention [76]. The Abecedarian (ABC) Project is one such intervention that reduced the risk of educational failure; children in the ABC Program were less likely to fail a grade, more likely to complete high school, and more likely to attend college [77]. Children who got the ABC intervention at an early age out-performed their peers at age 21. The full-scale IQ of those who got intervention was 94 at age 21, whereas the IQ of controls was 89 [77]. Roughly, 36% of adults who got the ABC intervention attended a 4-year college, while just 14% of control adults attended college ($p < 0.01$). These findings suggest that much of the CI that results from poverty could be prevented.

## Can CI in General Be Prevented?

Can CI be remediated, if people receive aggressive medical intervention? Can we mitigate CI after it has already occurred? These questions will form the focus for much of the rest of the book. We will explore these questions in the light of those few conditions where mitigation has already been tried. However, we must bear in mind that most efforts at cognitive remediation have been done in adults, not in children, even though children have more years to live and thus more to lose. It is ironic that the best-researched language remediation is for stroke, a medical condition largely limited to adults at the end of life, rather than to children at the beginning of life.

In the United States, where first-rate medical care is rationed on the basis of who can pay, it has been estimated that up to 17% of all current cases of mental retardation were preventable [78]. About 7.5 million people in the United States have mental retardation (IQ < 70), so this suggests that roughly 1.3 million Americans have a preventable form of retardation, while a far larger number of Americans may have a preventable form of CI.

Is CI from treated medical illness decreasing fast enough to account for the rise in IQ in the United States? Unfortunately, we do not have sufficient data to determine with surety if CI from medical illness can fully account for the progressive rise in IQ known as the Flynn effect [79–81]. Nevertheless, we have identified 30 conditions, some of which affect more than 10 million young people in the United States, and some of which reduce IQ by more than 20 points.

Average IQ loss for these medical conditions is 9.0 points, and crude calculation suggests that 389 million IQ points have potentially been lost. If 74 million young people are at risk, this amounts to 5.3 IQ points per person. Nevertheless, this calculation is sure to be flawed, because it assumes that IQ loss is spread equally among all American children, and because it assumes that existing public health measures have not prevented any potential IQ loss.

Loss of verbal IQ is unlikely to affect all children equally. Those children who are most disadvantaged by poverty will likely bear the greatest burden of CI, while wealthier children will be better able to realize their potential. In short, the loss of IQ points will be such that the gap between the "haves" and the "have-nots" is likely to grow wider with time.

It is also likely that much of the potential IQ loss has not happened, because of the success of public health interventions. The reduction of environmental lead is perhaps the greatest public health triumph in the United States in the last century. Similarly, the incidence of FAS has fallen by 74% in the United States over the past decade. Medicines used to treat pregnant mothers with AIDS are becoming more affordable; thus, far fewer children are born with HIV. And levels of PAH, arsenic, PCB, and mercury in the environment continue to fall, though air pollution is still a problem. Without doubt, some potential loss of IQ has been averted through existing public health measures.

Nevertheless, it is at least plausible that the ongoing rise in IQ points – what we know as the Flynn effect – results from a general improvement in public health in the United States. Hence, further mitigation of public health problems could potentially result in a greater increase in IQ in the United States.

A great many questions remain to be answered. Are there additional medical conditions – perhaps quite common in the United States – that reduce IQ, but are not yet proven to do so? For example, we do not know if childhood obesity, lack of physical exercise, vitamin D deficiency, or chronic ear infection has an impact on cognition in the United States. Is it possible that IQ is significantly or substantially lower only if children suffer multiple setbacks? For example, does CI happen solely in the context of poverty or are wealthy children also at risk? Would preventing malnutrition perhaps increase a child's cognitive resilience, so that other medical storms might be weathered more successfully? Or are some medical conditions so devastating that they produce permanent impairment, even if intervention is tried? What is the most effective way to mitigate the damage done by common causes of CI? Is it overly optimistic to imagine that there is a solution to the problem of impaired cognition?

# Chapter 10
# Impact of Medical Conditions on Human IQ Worldwide

We have discussed a range of medical problems that can potentially reduce IQ in the United States in Chap. 9. The estimated average IQ loss per condition in the United States is 9.0 points, and the aggregate number of IQ points lost to a remediable medical challenge is about 389 million points. Assuming that there are 74 million young people in the United States, an average of more than 5 IQ points are potentially at risk per American child. This potential IQ loss could happen despite the availability of high-quality medical care to those who can afford it.

## What Medical Challenges Depress IQ Worldwide?

It seems likely that, in the absence of high-quality medical care, the impact of the medical environment on cognition would be far more damaging. If language impairment and IQ decrements are truly a common consequence of various medical problems, then the absence of effective treatment for those problems would aggravate the outcome. Here, we provide a brief catalog of illnesses, syndromes, and diseases that may be common enough and damaging enough to reduce IQ worldwide, to at least some degree. As before, this list is not exhaustive, but rather it represents a distillation of research, addressing the relationship between medical environment and language impairment.

The following list of medical conditions that potentially depress IQ worldwide was compiled using just two criteria: the condition has been shown to cause a deficit in verbal or full-scale IQ; and the condition is preventable, curable, or treatable. As before, we include specifically those medical conditions and illnesses that afflict children, and we exclude medical conditions for which no treatment exists. The following list is arranged in an approximate order of the number of children affected, from most to least (Table 10.1). We note that some of these conditions have already been discussed, and hence will be noted here only briefly. Incidence estimates are based on an assumption that there are about 6.6 billion people in the world, with approximately 2 billion of them under age 15 [1]:

R.G. Steen, *Human Intelligence and Medical Illness*, The Springer Series on Human Exceptionality, DOI 10.1007/978-1-4419-0092-0_10,
© Springer Science+Business Media, LLC 2009

**Table 10.1** "Maximum IQ points at risk" was calculated as (People at risk) × (Estimated IQ risk per condition). This is done so as to give an idea of the scope of the problem, ignoring the fact that some people are affected by multiple medical conditions and that these conditions may have additive or synergistic effects. These numbers are estimates only, as both the number of people at risk and the VIQ loss are approximated. Poverty is not included in this list, but is discussed in Chaps. 7 and 10. In some cases, "Estimated people at risk" was calculated as people at risk per year times 20 years (the duration of a generation)

|  | Medical condition | Estimated people at risk | Estimated IQ risk per condition | Maximum IQ points at risk | Prevalence reference | IQ loss reference |
|---|---|---|---|---|---|---|
| 1 | Malaria | 2,000 million | −23 | 46,000,000,000 | [2] | [4] |
| 2 | Iodine deficiency | 1,990 million | −14 | 28,000,000,000 | [6] | [6] |
| 3 | Iron deficiency | 2,000 million | −9 | 18,000,000,000 | [9] | [10] |
| 4 | Hookworm infestation | 740 million | −15 | 11,000,000,000 | [13] | [15] |
| 5 | Roundworm infection | 1,200 million | −4 | 4,800,000,000 | [16] | [19] |
| 6 | Snail fever | 200 million | −19 | 3,800,000,000 | [20] | [21] |
| 7 | Preterm birth | 330 million | −11 | 3,600,000,000 | [23] | [25] |
| 8 | Attention deficit/hyperactivity | 330 million | −10 | 3,300,000,000 | [26] | [28] |
| 9 | Congenital toxoplasmosis | 260 million | −12 | 3,100,000,000 | [30] | [31] |
| 10 | Post-traumatic stress disorder | 330 million | −8 | 2,600,000,000 | [32] | [33] |
| 11 | Childhood neglect | 200 million | −8 | 1,600,000,000 | [35] | [37] |
| 12 | Whipworm infection | 795 million | −2 | 1,600,000,000 | [16] | [38] |
| 13 | Malnutrition/starvation | 149 million | −9 | 1,300,000,000 | [39] | [41] |
| 14 | Rubella infection | 100 million | −13 | 1,300,000,000 | [43] | [43] |
| 15 | Air pollution | 100 million | −12 | 1,200,000,000 | [44] | [46] |
| 16 | Diabetes | 200 million | −6 | 1,200,000,000 | [49] | [52] |
| 17 | Childhood cigarette smoking | 180 million | −6 | 1,080,000,000 | [56] | [57] |
| 18 | Depression | 330 million | −3 | 990,000,000 | [32] | [59] |
| 19 | Neurocysticercosis | 50 million | −17 | 850,000,000 | [60] | [61] |
| 20 | Lead exposure (>10 μg/dL) | 120 million | −7 | 840,000,000 | [62] | [63] |
| 21 | Early infant weaning to formula | 200 million | −4 | 800,000,000 | [66] | [66] |
| 22 | Childhood abuse | 70 million | −8 | 560,000,000 | [35] | [67] |

| | | | | | |
|---|---|---|---|---|---|
| 23 | Cerebral palsy/birth asphyxia | 16 million | −33 | 520,000,000 | [68] | [69] |
| 24 | Human immunodeficiency virus | 30 million | −15 | 450,000,000 | NY Times, 2007 | [70] |
| 25 | Obstructive sleep apnea | 40 million | −10 | 400,000,000 | [72] | [73] |
| 26 | Giardia infection | 100 million | −4 | 400,000,000 | Textbook estimate | [76] |
| 27 | Elephantiasis (lymphatic filiarisis) | 120 million | −3 | 360,000,000 | [16] | No data-estimated |
| 28 | Asthma | 120 million | −3 | 360,000,000 | [80] | [81] |
| 29 | Allergic rhinitis | 100 million | −3 | 300,000,000 | [83] | [20] |
| 30 | Arsenic poisoning | 50 million | −6 | 300,000,000 | [87] | [88] |
| 31 | Congenital hearing loss | 7 million | −27 | 189,000,000 | [90] | [92] |
| 32 | Epilepsy | 15 million | −12 | 180,000,000 | [96] | [98] |
| 33 | Very low birth weight | 21 million | −7 | 147,000,000 | [99] | [100] |
| 34 | PCB exposure | 15 million | −6 | 90,000,000 | [102] | [103] |
| 35 | River blindness | 37 million | −2 | 74,000,000 | [16] | [109] |
| 36 | Sickle cell disease | 6 million | −12 | 72,000,000 | [110] | [112] |
| 37 | Mercury pollution | 24 million | −3 | 72,000,000 | [113] | [114] |
| 38 | Congenital hypopituitarism | 3 million | −23 | 70,000,000 | [116] | [117] |
| 39 | Anxiety disorder | 15 million | −4 | 60,000,000 | [118] | [119] |
| 40 | Dengue fever | 20 million | −3 | 60,000,000 | [121] | No data-estimated |
| 41 | Traumatic brain injury | 4 million | −9 | 36,000,000 | [96] | [125] |
| 42 | Bipolar disorder | 3 million | −10 | 30,000,000 | [149] | [128] |
| 43 | Chagas disease | 9 million | −3 | 27,000,000 | [16] | [130] |
| 44 | Childhood tuberculous meningitis | 600,000 | −28 | 16,800,000 | [132] | [133] |
| 45 | Kernicterus | 700,000 | −18 | 12,600,000 | [134] | [135] |
| 46 | Pneumococcal meningitis | 3 million | −4 | 12,000,000 | [138] | [140] |
| 47 | Viral encephalitis | 1 million | −12 | 12,000,000 | [142] | [150] |
| 48 | Fetal alcohol syndrome | 1 million | −7 | 7,000,000 | [144] | [55] |
| 49 | Congenital heart defects | 2 million | −2 | 4,000,000 | [110] | [145] |

(continued)

**Table 10.1** (continued)

| | Medical condition | Estimated people at risk | Estimated IQ risk per condition | Maximum IQ points at risk | Prevalence reference | IQ loss reference |
|---|---|---|---|---|---|---|
| 50 | Guinea worm disease | 1 million | −3 | 3,000,000 | [146] | No data-estimated |
| | | **More than 2 billion people at risk** | **Mean IQ loss = −9.8** | **141.8 billion IQ points at risk** | | |

**Assumptions made:**

Cognitive effects of an illness are assumed to be chronic, even if the illness itself is curable

All calculations are based on global prevalence (# existing cases of a disease in the world at a given time)

When necessary, global prevalence was calculated as an average yearly incidence times width of the age class

Width of the age class is assumed to be 20 years

When global prevalence numbers are unavailable, we used national prevalence numbers from a heavily impacted country. Verbal IQ numbers are used when available, full scale IQ numbers are the second choice

When IQ scores are unavailable, the ratio of patient:control scores has been used to approximate the IQ

When no IQ data are available but a cognitive deficit is likely, a 3 point deficit is assumed

1. *Malaria*: About 2 billion people worldwide are exposed each year to malaria, resulting in 1 million deaths, and most of the deaths affect children under age 5 in sub-Saharan Africa [2]. Malaria caused a 15% deficit in language and mathematical scores in Sri Lankan school children in one study [3] and a 23% deficit in language scores in a second Sri Lankan study [4]. Intervention to treat malaria resulted in a 26% increase in language and mathematics scores among 295 Sri Lankan children after just 9 months [4]. Yet a simple insecticide-treated mosquito net around a sleeping area can substantially reduce the risk of malaria. Mosquito netting around a pregnant woman reduced the risk of low-birth weight infants by 23%, the rate of miscarriage by 33%, and malarial parasite infestation of the placenta by 23% [5].

2. *Iodine deficiency*: Nearly 2 billion people worldwide have insufficient iodine intake [6], and women who become pregnant while they are iodine deficient can bear severely retarded children. Several large studies show that IQ scores average about 14 points lower among children and adolescents with iodine deficiency [6]. Women who eat seafood during pregnancy supply the growing fetus with iodine and other trace nutrients that are largely unavailable from other sources, and thereby decrease the risk of having an impaired child. Among 11,875 women, those with the lowest seafood consumption had a 48% higher risk of bearing children with a low verbal IQ at age 8 [7]. In addition, inherited abnormalities of iodine metabolism adversely impact children, causing congenital hypothyroidism. Even if treated, congenital hypothyroidism can result in CI; in one study, verbal IQ in patients was 7 points lower than in healthy siblings [8].

3. *Iron deficiency*: Nutritional iron deficiency is another common dietary problem, affecting up to 2 billion people worldwide [9]. Iron deficiency is especially common in countries where women wean infants to artificial formula, which often lacks sufficient iron. Teenagers who were identified as being iron deficient during infancy had a full-scale IQ that was about 9 points lower than healthy matched children at age 19, and this gap widened to 25 points if chronic iron deficiency co-occurred with poverty [10]. Low prenatal iron availability is also associated with a threefold higher risk of mental retardation [11]. Among pregnant women in Tanzania at risk of iron-deficiency anemia, iron supplements led to a significant reduction in the number of children born small for gestational age [12]. Whether iron supplementation would also result in improved cognition among children is unknown, but it is a reasonable hypothesis.

4. *Hookworm infestation*: Parasitic worms infest more than one-quarter of the world's population. The human hookworm (*Ancylostoma* or *Necator*) probably infects 740 million people worldwide, as it is epidemic in parts of South America, Africa, and southern Asia [13]. Hookworms cause intestinal hemorrhage, with severe anemia, chronic diarrhea, protein deficiency, and food malabsorption. Since hookworm infestation is more prevalent among people who live in extreme poverty, this can mean that less nutritional benefit is obtained from food that is already scarce. In one area of Brazil, 68% of 1,332 people tested had *Necator* infestation, and the prevalence of anemia from

infestation was 12% [14]. Among 432 children tested for hookworm in Indonesia, infestation explained lower scores in verbal fluency, memory, concentration, and reasoning [15].

5. *Roundworm infection*: Roundworms (*Ascaris*) live in the soil and can infect people who have had an accidental contact with eggs or larvae. Roughly, 1.2 billion people in the world are infected by *Ascaris* – even though worm infestations are easily cured – because deworming medicine is prohibitively expensive in many parts of the world [16]. The gut of a child living in poverty is often parasitized by several types of parasitic worms; in one district in Kenya, 92% of children were infected with *Ascaris*, whipworm, hookworm, or snail fever, and most children were infected with two or more gut parasites [17]. Among 319 schoolchildren in the Philippines, the prevalence of *Ascaris* was 74% [18]. Treatment of *Ascaris* infection in primary school children in Indonesia was associated with a 4-point improvement in performance on a well-accepted measure of cognitive ability [19]. This striking improvement in cognition was realized within just 5 months of deworming treatment.

6. *Snail fever*: Snail fever is caused by schistosomes – parasitic organisms also known as blood flukes – that live in infected fresh water and that, upon infection of humans, localize to small blood vessels, especially in the gut. These parasites are epidemic in Asia, Africa, the Middle East, South America, and the Caribbean, and at least 200 million people in 74 countries are infected, and 20 million have severe illness [20]. Treatment for schistosome infection in Chinese school children resulted in a 19% improvement in verbal fluency in 3 months, while the untreated group actually performed worse over time [21]. In treated children (5–7 years old), fluency scores increased by 24%, while untreated children suffered a slight decline in fluency, perhaps because their infections remained untreated. In Tanzanian school children with severe schisosomiasis, short-term verbal memory and reaction time were impaired [22]. These findings do not necessarily mean that there is a brain damage in untreated children, but it is difficult for students to do their best on a test when they have dysentery, painful liver enlargement, skin rash, and chronic anemia.

7. *Preterm birth*: Preterm birth has been estimated to affect 5% of births in developed nations and up to 25% of births in developing nations [23], but we will assume here that preventable prematurity – which results from inadequate prenatal care – affects 5% of all births worldwide. Premature children have a smaller brain than normal, even 8 years later, and brain size is a significant predictor of verbal IQ [24]. In a meta-analysis of 15 studies which included 1,556 cases and 1,720 controls, premature birth was associated with an 11-point deficit in cognitive test scores in school-age children [25].

8. *Attention-deficit/hyperactivity disorder (ADHD)*: Prevalence of ADHD is estimated to be 3–7% of all children [26]; while we do not know for certain that ADHD prevalence is equivalent in the developed and the developing world, we will assume that the worldwide prevalence is 5%. Adolescents and adults with ADHD usually show a deficit in short-term verbal memory [27]. Verbal IQ in children with ADHD is as much as 20 points lower than that of matched healthy children [28], but we assume that a 10-point deficit in verbal IQ is representative.

9. *Congenital toxoplasmosis*: *Toxoplasmosa gondii* is an obligate intracellular parasite that is transmitted to humans by infected cats or undercooked meat, and it has been estimated to infect up to a third of the world's population [29], or about 2 billion people. However, only about 13% of affected people show clinical signs of intracranial infection [30], so we calculate that roughly 260 million people will show CI from toxoplasmosis. Among children impaired by infection – which may include many children lacking access to health care – verbal IQ score was roughly 12 points below normal [31].

10. *Post-traumatic stress disorder (PTSD)*: In a war-torn and disaster-sundered world, PTSD is far more common than it should be; the lifetime prevalence of PTSD is estimated at 10% in women and 5% in men [32]. If we assume that 5% of people have PTSD, this amounts to 330 million people overall. Among 59 children who were diagnosed with PTSD after exposure to childhood violence, the average verbal IQ was 92 [33]. Among adult women who had PTSD as a result of childhood sexual abuse, there was a deficit specifically in verbal memory, without a corresponding deficit in full-scale IQ [34].

11. *Childhood neglect*: A recent nationally representative survey in the United States concluded that 3% of all children suffer childhood neglect [35], which may mean that 200 million people worldwide have suffered neglect. Children who are physically stunted and emotionally neglected at age 2 years tend to achieve poorly on standardized tests of cognition at age 18 years [36]. When children who were abandoned shortly after birth and institutionalized in Romania were randomized to receive either continued institutional care or home foster care, there was a striking difference in IQ within 33 months [37]. Institutionalized children had an IQ of 73, whereas children randomized to foster care had an IQ of 81, roughly 8 points higher than children who suffered continued neglect. Because children were randomized to care in this study, this is a particularly strong demonstration that childhood neglect can be remedied by human contact.

12. *Whipworm infection*: Whipworms (*Trichuris*) also live in the soil and are transmitted to humans by contact with eggs or larvae. *Trichuris* infects nearly 800 million people worldwide [16], though deworming medicines are also effective for these parasites. Among 319 schoolchildren in the Philippines, the prevalence of *Trichuris* was 92% [18], and infection was associated with poor performance on tests of verbal fluency. Nevertheless, the increase in verbal ability exhibited by children after deworming is rather small, and often not statistically significant [38], hence we will assume a minimal change of 2 verbal IQ points post-treatment.

13. *Malnutrition/starvation*: The World Health Organization (WHO) has estimated that the global prevalence of stunting in preschool children was 149 million in 2005 [39]. Stunting is a severe form of malnutrition; children are not merely underweight, they are actually impaired in development by a lack of food. Boys born preterm who were at risk of stunting were randomized to get nutritional supplementation, as opposed to feeding with a standard formula; well-nourished children had a 12-point higher verbal IQ at age 8 [40]. Children with severe malnutrition at age 3 (e.g., kwashiorkor, anemia, sparse hair) showed a 9-point deficit in verbal IQ by age 11 [41].

14. *Rubella infection*: Rubella, also known as German measles, is usually a mild illness, but infection of a pregnant woman during the first trimester causes miscarriage or congenital infection of the newborn in up to 90% of cases [42]. Rubella affected perhaps 10% of all pregnant women prior to the introduction of an effective vaccine in 1969, and 30% of infants born to an infected mother had congenital infection [43], so roughly 3% of infants may have been impaired. The vaccine is still not given to all pregnant women, and many adults now were born when vaccination was not possible; thus, we estimate the worldwide prevalence of congenital rubella infection to be about 100 million. Introduction of the rubella vaccine in the United States reduced the incidence of rubella infection by 99%. But, without the vaccination – which is still unavailable in most of the world – congenital rubella can cause deafness, cataract, cardiac anomalies, and mental retardation. Among people with congenital rubella syndrome, mental retardation is present in 42% (full-scale IQ < 70), with hearing loss in 66%, and ocular disease in 78% [43].

15. *Air pollution*: China has been called the "air pollution capital of the world" and it has been estimated that 100 million people live in cities where toxins in the air reach levels considered "very dangerous," with the situation likely to worsen [44]. Air pollution takes many forms, but the forms considered here are carbon monoxide and polycyclic aromatic hydrocarbons. Carbon monoxide (CO) is an odorless and invisible gas produced as a by-product of incomplete combustion; it is the leading cause of poisoning injury and death worldwide [45]. Experimental (2 h) exposure to moderate CO poisoning by a residential kerosene stove resulted in an impairment of memory, attention, concentration, learning, visuo-motor skills, information processing, and abstract thinking, with verbal ability depressed 12% [46]. Polycyclic aromatic hydrocarbons (PAH) are released by burning fossil fuel, and they contribute to a delay in cognitive development of children. PAH levels, measured in the air near pregnant women by "personal air sampling" were able to predict which offspring would be developmentally delayed at 3 years of age [47]. In a cohort of 150 non-smoking women in China, PAH levels measured in cord blood of newborn infants predicted which infants would suffer growth delay at age 30 months [48]. Because these women did not smoke, the only source of PAH exposure was a nearby coal-fired power plant.

16. *Diabetes*: Roughly, 200 million people currently have diabetes, with this number projected to rise sharply over the next few decades, as obesity becomes more prevalent [49]. Well-controlled childhood diabetes may not produce any decrement in academic performance [50], but well-controlled diabetes requires ongoing high-quality medical care. In children with poorly controlled diabetes or a history of hypoglycemic seizures, there can be reduced verbal comprehension, with impairment in short-term and word memory [51]. Six years after onset of childhood diabetes, verbal IQ scores in patients were as much as 6 points lower than in healthy matched children, with a 5-point decrement in full-scale IQ [52]. Children born to obese mothers who are pre-diabetic also have a lower full-scale IQ [53]. Nevertheless, diabetes in adulthood may have no impact on cognition [54].

17. *Childhood cigarette smoking*: Maternal smoking during pregnancy reduces verbal reasoning ability and full-scale IQ of offspring by less than a point [55], but when children themselves smoke there can be serious cognitive consequences. Worldwide, about 9% of children aged 13 to 15 years smoke cigarettes [56], which amounts to 180 million children. Comparing non-smoking children to children who smoked at least half a pack per day, verbal IQ was 6 points lower and verbal comprehension was 7 points lower among children who smoke [57].

18. *Depression*: The burden of depression may be smaller in poor countries than in rich countries; depression is thought to affect roughly 1% of people in Africa, but 9% of people in high-income countries [32]. Yet other studies have sharply disagreed with this estimate; the 1-month prevalence of depression and anxiety in Zimbabwe was estimated to be 16% [58]. We assume that, each year, roughly 5–8% of adults are depressed, and we estimate that at least 330 million people worldwide suffer from depression with coincident impairment of cognition. Recent evidence suggests that patients who remit from depression enjoy about a 3 point increase in IQ, and improved cognitive function may actually be a marker for recovery from depression [59].

19. *Neurocysticercosis (NCC)*: Human infection with pork tapeworm occurs if pork is consumed without being properly cooked, and NCC results if tapeworm larvae invade the brain [60]. NCC is probably the most common parasitic brain infection, afflicting an estimated 50 million people worldwide and causing 50,000 deaths annually. NCC cysts have been reported in up to 2% of all autopsies in Mexico, and infection is also common in Central and South America, Africa, Southern Asia, China, and Eastern Europe. NCC is a major cause of epilepsy in the developing world, and it can also cause amnesia, apathy, emotional instability, and hallucinations [61]. Whether NCC causes verbal impairment is not known with certainty, but patients with NCC and epilepsy scored 17% lower in oral comprehension than did patients with epilepsy alone [61].

20. *Lead exposure*: Blood lead levels higher than 10 µg/dL are estimated to affect 120 million people worldwide [62]. Even at very low levels of lead exposure, there can be problems; full-scale IQ declined by about 7.4 points as the lifetime average blood lead increased from 1 to 10 µg/dL [63]. New results suggest that bone lead levels are a better measure of lead exposure over time than are blood lead levels, as blood levels decrease soon after exposure. For every doubling of bone lead levels, children lose about 4 verbal IQ points [64].

21. *Early infant weaning to formula*: Increased duration of nursing with breast milk is associated with an increase in verbal and full scale IQ. Considering low-birth weight babies, those who were breastfed for 8 months or longer had a verbal IQ about 10 points higher than children who did not receive breast milk [65]. In normal-weight babies, evidence suggests that breast milk feeding is associated with a 4-point increase in verbal IQ, if other potential confounders are controlled [66]. Nevertheless, about 10% of infants in developed countries are probably not breast fed [66]. We, therefore, estimate that about 200 million people show at least some degree of CI as a function of early weaning.

22. *Childhood abuse*: A recent nationally representative survey in the United States concluded that 10% of children suffer physical abuse [35], suggesting that up to 700 million children in the world may be abused. Exposure to domestic violence reduces IQ; this conclusion is based on a study of 1,116 twins, so it is probably a robust estimate [67]. Relative to children who report no domestic violence, children who are exposed to low levels of domestic violence show a reduction of 1 point in IQ, and children exposed to high levels of such violence show an 8 point IQ reduction. We will assume that only 10% of abused children suffer abuse serious enough to result in an 8 point IQ decline.

23. *Cerebral palsy/birth asphyxia*: Cerebral palsy is a form of birth injury caused by severe hypoxia (lack of oxygen), and it is associated with poor medical care in the prenatal or perinatal period. The overall prevalence of cerebral palsy in children aged 3–10 years is 2.4 per 1,000 children [68], which suggests that up to 24 million people in the world have cerebral palsy at birth. In a random sample of 68 children with cerebral palsy, average full-scale IQ was just 67 [69].

24. *Human immunodeficiency virus (HIV) infection*: The *New York Times* recently reported that 33 million people currently have HIV infection. Children with clinically stable infection with HIV, the virus that causes AIDS, had a verbal IQ of 82 to 85, substantially below that of healthy children [70]. Yet transmission of AIDS from mother to child can be prevented in 85% of HIV-infected mothers, even if children are breastfed, provided the mothers are given antiviral treatment [71]. Childhood AIDS is, therefore, a preventable cause of CI.

25. *Obstructive sleep apnea*: Chronic obstructive sleep apnea (OSA) is a cessation of breathing that occurs during sleep, and prevalence is approximately 2% in children [72]. Children with severe OSA showed a 15-point decrement in full-scale IQ, with specific problems in verbal and working memory, compared to healthy matched children [73]. These deficits in IQ and executive function are well documented, although in a small group of children with severe OSA. It is not known whether mild apnea or snoring is associated with similar deficits. We will assume that a typical level of sleep apnea results in a 10-point decrement in verbal IQ, at least in part because of daytime sleepiness.

26. *Giardia infection*: *Giardia* – an intestinal microbe that can cause chronic diarrhea and malaise – is widely believed to affect 100 million people. Children with a heavy burden of early childhood diarrhea from *Giardia* show a deficit in semantic fluency at age 6 to 12 years [74]. This deficit remains after statistically adjusting for maternal education, breastfeeding, and schooling. Even 4 to 7 years later, children who suffer severe diarrhea in early childhood are verbally impaired with respect to children free of chronic diarrhea [75]. The reported reduction in IQ scores among children with more than a single episode of *Giardia*-related diarrhea is 4 points [76].

27. *Elephantiasis (lymphatic filariasis)*: Elephantiasis is a condition resulting from a massive infestation of tiny parasitic worms that obstruct lymph vessels, so that lymph fluid cannot drain away from extremities. Parasites are spread by mosquito bite. Heavy infestation can lead to grotesque swelling of the legs, arms, breasts, or genitalia, and extremities can increase in size dramatically.

Filarial infestation is estimated to affect up to 120 million people worldwide [16]. In patients with lymphatic filariasis in Kenya, 55% of adult male patients had swelling of the testicles, and 9% of patients had lymphedema of the leg – the classic manifestation of elephantiasis [77]. The WHO considers elephantiasis to be the fourth leading cause of permanent disability. It is well documented that epidemic filarial infection leads to a serious loss of economic productivity due to incapacitation of workers [78]. Recent clinical evidence suggests that infection typically starts in childhood and that the pathogenesis of disease is not solely a result of obstruction of lymph vessels; release of toxins by the worms and co-infection with bacteria may account for some disease morbidity [79]. It is not known for certain if lymphatic filariasis results in CI, but we will assume that there is some impairment because of the lost productivity and general malaise that accompanies serious illness.

28. *Asthma*: While the diagnosis of asthma is somewhat problematic, a clinical diagnosis is typically made in about 10% of children [80], which amounts to 600 million people worldwide. We will assume that only 20% of these people – 120 million people in total – suffer CI from asthma. Children with asthma serious enough to lead to hospitalization show a 3-point decrement in full-scale IQ, as compared to asthma patients who are not hospitalized [81], though the extent of deficit compared to similar healthy children is unknown. Asthmatic patients show a decrement in attention, but it is not known whether and to what extent attention problems are associated with deficits in verbal ability [82]. We will assume that there is only a 3-point verbal IQ deficit in severe asthma.

29. *Allergic rhinitis (AR)*: Allergic rhinitis affects roughly 5% of people worldwide; since it is treated with prescription medication in just 12% of cases in the United States, we will assume that it is predominantly untreated in the rest of the world [83]. In patients who are not medicated, CI results from general malaise, insufficient nighttime sleep, and daytime sleepiness [84], as well as impairment in working memory, psychomotor speed, reasoning, computation, and vigilance [85]. We will assume that AR is associated with a degree of CI comparable to asthma.

30. *Arsenic poisoning*: Arsenicosis typically results from exposure to hazardous waste or contaminated well water, and high levels of arsenic have been found in wells in Argentina, Canada, China, India, and Mexico, as well as the United States [86]. Testing of 21,155 wells in China found that 5% of wells showed excessive levels of arsenic in groundwater [87]; if this is true of China as a whole, then at least 50 million Chinese have arsenicosis. Childhood exposure to high levels of arsenic can reduce verbal IQ by 6 points, other things being equal [88]. There is a dose-response relationship such that children exposed to high levels of arsenic show more impairment of verbal ability [89].

31. *Congenital hearing loss*: Congenital bilateral permanent hearing loss affects roughly 112 infants for every 100,000 births [90], which translates to 7 million people worldwide. Hearing impairment is associated with deficits specifically in verbal IQ, and there can also be problems with speech articulation and language learning [91]. Scores in language ability were 27 points lower in children with

profound hearing loss, compared to children mildly impaired by hearing loss [92], which may approximate the deficit expected if hearing loss is uncorrected. In a study of 122 deaf pupils in the Netherlands, preventable causes of deafness were more common than inherited causes, and every deaf child had moderate-to-severe retardation [93]. It is possible that, if testing is done so as not to discriminate against hearing disability, the IQ of deaf children would be comparable to children with normal hearing [94], yet the fact remains that life itself discriminates against the deaf, and the cost of a hearing aid can be prohibitively high. Early hearing correction is associated with improved scores on language comprehension [95].

32. *Epilepsy*: The lifetime prevalence of epilepsy in the United Kingdom has been estimated at 4 per 1,000 of the population, and epilepsy prevalence is almost certainly higher in the developing world [96]. This suggests that there are at least 26 million people with epilepsy worldwide. Among 1,017 children with epilepsy in South Africa, there was no evidence of brain damage or defect in 57% of cases [97]. Epilepsy without brain injury often results from preventable causes, such as meningitis, tuberculosis, or NCC infection, so we estimate that 57% of epilepsy cases worldwide could have been prevented. Children with epilepsy score lower than healthy matched children on all cognitive measures, with 12 point lower scores on verbal IQ [98].

33. *Very low birthweight (VLBW)*: The United Nations Children's Fund has estimated that more than 20 million infants globally are born at a birthweight less than 2,500 grams [99], which is defined as VLBW. Children with VLBW have striking deficits in cognitive ability compared to normal birthweight children [100]. Overall, VLBW children show a 7-point deficit in verbal IQ, a 10-point deficit in full-scale IQ, and a 12-point deficit in performance IQ, and there are also deficits in mathematics and gross motor skills.

34. *PCB exposure*: Polychlorinated biphenyls (PCBs) are organic chlorine chemicals that were once widely used in industry, but their use has been curtailed because PCBs can accumulate in the environment and in living tissues [101]. A national survey completed in the United States in 2002 suggests that roughly 5% of Americans have elevated body levels of PCB [102]. We will assume that PCB pollution is limited to the United States, so we estimate that 15 million people worldwide suffer from PCB exposure. There is evidence that prenatal exposure to PCBs causes significant reduction in verbal IQ and full-scale IQ with the strongest effects on memory and attention. IQ scores of the most highly exposed children were more than 6 points lower than those of healthy children, and PCBs reduce scores on tests of vocabulary, information, and similarities [103]. In New York State, adolescents exposed to PCB pollution showed an impairment of long-term memory, reading comprehension, and general knowledge [104]. In Dutch children exposed to PCBs and dioxins in mildly polluted water, *in utero* exposure to PCBs was associated with poor cognitive function in preschool [105]. Yet other studies have been unable to replicate the finding that background PCB levels are harmful, hence these results are controversial [106].

35. *River blindness (Onchocerciasis)*: It has been estimated that 37 million people in the world have been exposed to the parasite that causes river blindness [16]. Onchocerciasis results from a worm infestation, and the parasite spreads from one person to the other by biting black flies. Infestation typically produces skin and eye lesions, and results in blindness in 1% of patients [107]. Onchocerciasis has been linked to an increased risk of epilepsy, but this relationship is not proven [108]. Nevertheless, adult-onset blindness is disabling; among 6,234 adults who became legally blind, cognitive test scores declined by about 2 IQ points [109].

36. *Sickle cell disease (SCD)*: The March of Dimes has estimated that 300,000 children are born each year with SCD [110], which would translate to 6 million people per generation. Cognitive function of children with SCD is impaired, even among children free of stroke [111]. In 54 children with SCD, verbal IQ was 12 points lower than healthy children of the same age, race, and gender [112].

37. *Mercury pollution*: Mercury is released by coal-fired power plants, by gold mining, and by waste incineration, and it often contaminates seafood. Data from the National Health and Nutrition Examination Survey in the United States shows that 8% of American women have blood mercury levels higher than recommended by the Environmental Protection Agency [113]. This suggests that 24 million Americans are at risk of mercury poisoning, with less mercury toxicity elsewhere in the world, because industrial pollution is a bigger problem in the United States. At the highest levels of contamination found in the United States, it is likely that no more than about 3 IQ points are lost to mercury toxicity [114]. Though severe mercury poisoning causes a devastating form of retardation, evidence suggests that the amount of mercury present in dietary fish is responsible for roughly a 1-point IQ deficit [115].

38. *Congenital hypopituitarism*: The prevalence of hypopituitarism, due to insufficiency of the pituitary gland, is estimated to be 46 cases per 100,000 people [116], or about 3 million people worldwide. Clinical symptoms of illness are non-specific, with chronic fatigue, anorexia, weight loss, cold intolerance, hair loss, and cognitive slowing. However, the degree of cognitive slowing is striking; in a study of 10 children, all of whom had treated congenital hypopituitarism, verbal IQ was 23 points less than normal [117]. This deficit could arise through abnormal brain development during infancy, or it could be related to the fact that patients with chronic hypopituitarism can suffer low blood glucose or low growth hormone availability in infancy.

39. *Anxiety disorder*: The lifetime prevalence of generalized anxiety disorder in the United States is 5% [118], suggesting that up to 15 million Americans have the condition. Prevalence rates throughout the rest of the world – especially in the developing world – are unknown, so we will make a very conservative assumption that only Americans are affected. Comparison of the full-scale IQ of 57 children with an anxiety disorder to 103 matched children, free of anxiety, showed a 4-point IQ deficit in anxious children [119].

40. *Dengue fever*: This severe hemorrhagic fever, also called "breakbone fever," is caused by a mosquito-borne virus, and there are signs of central nervous system

involvement in 21% of patients [120]. There are an estimated 100 million cases per year worldwide and serious neurologic complications occur in about 1% of patients, including coma, stupor, convulsions, or behavioral disturbance [121]. This may mean that a million patients per year have reduced cognition as a result of dengue fever, and 20 million people in the age class up to 20 years may be affected. What proportion of these patients suffer permanent CI is not known, but patients with neurological problems at diagnosis can suffer problems months later. In India, more than half of all patients with neurologic manifestations of dengue show abnormal slowing of brain activity with potential permanent injury [122]. There can also be post-infectious fatigue, with incapacitating weariness months after the acute phase of illness resolves [123]. Among 127 patients admitted to hospital, 24% still had fatigue 2 months after release, which could potentially reduce their cognitive performance. We provisionally estimate that dengue fever reduces verbal IQ by 3 points.

41. *Traumatic brain injury (TBI)*: A recent epidemiological study of severe brain injury in the United Kingdom estimated that there are 60 cases per 100,000 people [96], which translates to about 4 million cases worldwide. Roughly, 23% of emergency department visits in the United States result from TBI with a loss of consciousness [124], though it is not known what proportion of these injuries result in CI. When children with TBI return to school, two-thirds of them have difficulties with school work, and there can be problems with memory and attention. Among 56 brain-injured children, average verbal IQ score was 91, well below normal [125]. Among children brain-injured as infants, the results were much worse; 48% of such children had an IQ below the 10th percentile, and the risk of poor academic performance among the brain-injured was 18-fold higher than normal [126]. Verbal learning scores can decline up to 14 years after severe TBI [127].

42. *Bipolar disorder (BD)*: The lifetime prevalence of BD in the United States is estimated to be about 1% [118], suggesting that roughly 3 million Americans suffer BD. A small study [128] identified a verbal IQ deficit of 22 points in BD patients, although these were patients at a state psychiatric hospital, so they likely had more serious impairment. We estimate that the average BD patient has a verbal IQ deficit of 10 points.

43. *Chagas disease*: Parasitic infestation with the human parasite *Trypanosoma cruzi* causes Chagas disease, and the parasite is passed from one person to an another by the bite of the triatomine or "kissing" bug. Chagas disease is epidemic throughout the Central and South America, where it is estimated to infect about 9 million people [129]. There are no recent references on CI as a function of the Chagas infection, but there is an old reference that reports significant deficits in reasoning, problem solving, learning, and information processing [130]. In the absence of recent data, we will assume a 3 point deficit in verbal IQ.

44. *Childhood tuberculous meningitis*: There were an estimated 8.9 million new cases of tuberculosis in 2004, and tuberculosis is the leading cause of death from a curable infectious disease [131]. Vaccination could prevent up to 30,000 cases of tuberculous meningitis per year [132]. We estimate that the

number of people at risk of CI is equivalent to 20 years of preventable tuberculous meningitis cases or 600,000 patients. Long-term follow up of 74 children who had childhood tuberculous meningitis shows that the median IQ was 72, or 28 points below normal [133].

45. *Kernicterus*: Neonatal jaundice is a very common condition, affecting up to 11% of term infants [134], so perhaps 700 million people in the world have had births complicated with some hyperbilirubinemia. Prior to birth, the placenta clears bilirubin – a product of hemoglobin breakdown – from the fetus, but after birth the infant must detoxify it. As the neonatal liver matures, jaundice resolves, but brain injury can result if bilirubin accumulates before it is excreted. Neonatal jaundice is routinely and successfully treated with exposure to bright lights [135], and few people realize how devastating it can be without treatment. Untreated hyperbilirubinemia can progress to kernicterus – characterized by cerebral palsy, hearing loss, and other neurologic problems – with 10% mortality and 70% long-term morbidity [136]. If just 0.1% of untreated hyperbilirubinemia cases progress to kernicterus, then 700,000 people have CI as a result of bilirubin accumulation. There are no reliable studies of CI from untreated hyperbilirubinema. However, in Zimbabwe, where the dark skin of infants makes it hard to diagnose and treat the condition, 26% of infants with extreme hyperbilirubinemia had CI at 1 year of age and 12% developed cerebral palsy [137]. Among children with a severe form of hyperbilirubinemia, there was an 18-point decrement in verbal IQ [135].

46. *Pneumococcal and other meningitis*: The "meningitis belt" of sub-Saharan Africa stretches from Senegal to Ethiopia, with an estimated population of 300 million, and waves of bacterial meningitis sweep through the area every 8–12 years. Each epidemic sickens from 300,000 to 3 million people [138], so we will assume that 3 million people are exposed per generation. Among children who develop bacterial meningitis, language difficulties are common, and young age at infection is a particularly strong risk factor [139]. Even in adults with bacterial meningitis, CI can be a problem. Among Dutch adults with meningococcal meningitis, cognitive slowing was present 2 years post-illness, and 27% of patients had a cognitive disorder [140]. Verbal reasoning scores were 4 points lower than in healthy matched subjects, though the sample size was too small for this difference to be statistically significant.

47. *Viral encephalitis*: Because so many viruses cause viral encephalitis [141], it is essentially impossible to know the prevalence of viral encephalitis. However, the annual incidence of enterovirus 71 (EV71) infection in Taiwan averages 4 per 100,000 [142]; if this annual incidence is true for the billion-strong Chinese population, then 800,000 Chinese are at risk of EV71 infection per generation. We will assume that, including other causes of viral encephalitis, the actual prevalence of viral encephalitis is 1 million per generation. A group of 142 Taiwanese children was monitored for 3 years after EV71 infection, and their cognitive ability was compared to healthy children. Among the most severely impacted children, verbal IQ was reduced by 18 points [143]. There is evidence to suggest that cognitive deficits after encephalitis change with time.

48. *Fetal alcohol syndrome (FAS)*: FAS was coded as a discharge diagnosis in 17 per 100,000 births in the United States in 2002, which represents a 74% reduction in FAS among newborns over the past decade [144]. Nevertheless, even accepting the reduced rate as typical of births worldwide – which is a very optimistic assumption – there are likely to be 1 million people with FAS. This estimate is conservative because some parts of the world have a much higher FAS rate. As noted, alcohol exposure in the second trimester results in an approximate 7-point IQ deficit [55].

49. *Congenital heart defects*: The March of Dimes has estimated that 1 million children are born with congenital heart defects each year [110]. We will assume that only 10% of those children – in particular, those with the mildest defects – would survive without surgical treatment, which is equivalent to 2 million people per generation. We further assume that the IQ of untreated surviving children in the developing world is comparable to the IQ of treated children surviving in the developed world. A study of 243 children who had surgery for congenital heart disease concluded that their verbal IQ was 2 points less than normal [145].

50. *Guinea worm disease*: Guinea worm disease or dracunculiasis is a parasitic disease found in villages in sub-Saharan Africa that have unsafe drinking water. A global guinea worm eradication campaign, begun in 1981, has already reduced the number of newly infected people by 98%, from 3.5 million in 1986 to 55,000 in 2002, and new cases are now limited to Ghana, Nigeria, and Sudan [146]. Nevertheless, there are still about 1 million new cases per generation. A guinea worm is up to 1 m in length and can cause painful boils in the skin that swell and burst over several weeks, thereby releasing larvae to complete the life cycle. An infected person typically has two boils at one time, suffers pain for up to a year after the larvae emerge, and misses an average of 100 days of work. In Nigeria, where dracunculiasis was common until recently, 21% of students were infected and infected pupils missed up to a quarter of the school year, a rate of absence 10-fold higher than normal [147]. Whether school absence is associated with verbal impairment may never be known, as the disease is now rare, but it is reasonable to assume that some degree of verbal impairment occurs as a result of dracunculiasis.

Each of these 50 worldwide causes of CI should be seen in context; some of them are rare and have relatively little impact, but others are strikingly common and can potentially co-occur with other listed conditions, perhaps leading to detrimental additive or synergistic effects.

In order to calculate the IQ loss per person, we will make a back-of-the-envelope calculation based on a few key assumptions. First, we will assume that IQ loss is concentrated among the poorest and most unfortunate 40% of people worldwide, and that the remaining 60% of people are unaffected by the illnesses and conditions listed. Secondly, we will assume that 80% of the total potential CI has already been averted, because of ongoing public health measures. Even with these optimistic assumptions, 2.6 billion people worldwide may now be showing an average decrement of about *11 IQ points per person*.

This is a conservative estimate, because we have assumed that each person has only one illness and that there are no synergistic interactions between illnesses that depress IQ more than would happen for individual illnesses. It is also conservative in that we have assumed that these 50 tabulated causes of IQ loss are the only causes of CI. Clearly, if other illnesses also depress the IQ but have not been tabulated, then the total IQ points potentially lost could be higher than 141.8 billion.

According to the World Health Organization (WHO), 5–15% of all children in the world (ages 3–15 years) are cognitively impaired [148]. Given that there are about 2 billion children in the world [1], this suggests that 100 to 300 million children have some degree of CI. The WHO has estimated that 10 to 30 million children are severely retarded (IQ range, 25–39), and an additional 60 to 80 million are moderately retarded (IQ range, 40–54) as a result of a preventable cause [148]. An unknown but larger number of children are mildly retarded (IQ range, 55–69), and a vastly larger number of children have CI too mild to be called retardation (IQ range, 70–85). Yet it is not clear how much of this burden of impairment could potentially have been prevented if quality medical care was available to all children.

# Chapter 11
# Medical Interventions for Cognitive Impairment

We have proposed that the Flynn effect – the progressive rise in IQ that is well documented over the past several decades – may have arisen because of successful intervention into medical conditions that depressed intelligence in the past. In the United States alone, more than 10 million children are at risk from the effects of poverty, and the average number of IQ points potentially lost to poverty is about 4. This means that 41.7 million IQ points may be at risk from the effects of poverty. Across the 30 conditions identified as problematic in the United States, the average IQ loss is about 9 points. Hence, the aggregate number of IQ points lost to a remediable medical challenge in the United States is about 389 million points. If there are 74 million young people in the United States, then an average of *more than 5 IQ points are at risk per person.*

However awful this is, it was almost certainly worse in the past. A child blessed with genes that could confer intelligence will not actually have a high IQ unless that child has a fairly benign medical environment. In the absence of disease, children can fulfill the promise of their genes; but disease can prevent a child from achieving her full potential, no matter how favorable her genes. Genes determine the potential, whereas environment determines the actual. In a sense, genes determine the range through which environment can modify an individual. Favorable genes and a benign environment may enable a child to reach her full genetic potential. But favorable genes and a malign environment may provide no better outcome for a child than would unfavorable genes. This forces us to consider a question relevant to millions of children who have perhaps already suffered a poor outcome from poverty or other remediable medical event.

Can children who have already lost IQ points to illness or circumstance be helped? Is there any hope for cognitive remediation in a child whose verbal ability is impaired by poverty, low birth weight, or an untreated parasite infection? Or is the burden of cognitive impairment in such a child permanent? In short, can medical intervention increase human intelligence?

To determine whether cognitive remediation is possible, we must be as rigorous as possible. Hence, we will focus primarily on language, which is perhaps the best indicator of intelligence, because it is both a measure of problem-solving ability and a predictor of social aptitude. Our question is, can a public health intervention improve the average cognitive ability of our nation's children?

R.G. Steen, *Human Intelligence and Medical Illness*, The Springer Series on Human Exceptionality, DOI 10.1007/978-1-4419-0092-0_11,
© Springer Science+Business Media, LLC 2009

## Testing a Medical Intervention

To rigorously determine whether medical intervention can improve language use, we require that an accurate measure of verbal ability be used objectively. Thus, we will limit our attention, insofar as possible, to what are called "randomized, double-blind, placebo-controlled clinical trials." In a randomized trial, people are randomly assigned to receive either an intervention or a placebo, meaning that the composition of the treated and untreated groups should be the same at the outset. In a double-blind trial, neither patients nor investigators know which treatment has been given to which patient, so no one has a realistic expectation as to what the outcome should be. In a placebo-controlled trial, patients who are untreated receive a form of placebo that seems just like the real treatment but which is inactive and not expected to work. For the sake of simplicity, we will henceforth call any randomized, blinded, placebo-controlled clinical trial an "RBP" trial, and we will assume that RBP trials are always the gold standard used to assess whether an intervention has been successful.

It is crucial to have participants in a clinical trial blinded as to what treatment is received because the human mind is very good at generating an "expected" outcome, even in the absence of treatment. A recent review of 75 different RBP trials of antidepressants showed that roughly 30% of depressed patients given placebo reported improvement in their condition, whereas just 50% of patients who received an active medication reported benefit [1]. Therefore, antidepressants are only a little more effective than inactive placebo, and the incremental benefit from active medication is far smaller than we would like [2]. It is not clear whether this means that the current antidepressants are ineffective or that placebo is surprisingly effective. In either case, the patients' tendency to get a "real" benefit from a sham treatment is called the placebo effect.

Popular understanding of the placebo effect holds that it is the therapeutic benefit that results from taking a sugar pill, the active response to an inactive medication. Yet, a crucial element of placebo treatment is that the recipient must believe that he has received active therapy; no placebo effect is likely if a person does not anticipate some benefit. A person primed to show a placebo effect is, therefore, in a state of willing expectancy. Therapy offers hope, and hope itself is therapeutic; in fact, it has been argued that hope *is* the placebo effect. [2]. Hence, it is only when hope cannot influence the outcome of a clinical trial that we can be sure that the findings of that trial are real and not the result of a runaway placebo effect.

## Language Remediation after Stroke: A Proof of Principle

Stroke can result in overt and often extensive brain damage. A stroke occurs when the blood flow to the brain is blocked in some way, so that brain tissue becomes oxygen starved and dies. This is sometimes fatal and often devastating in adults, because the brain is vulnerable to hypoxia after even a short period of time with inadequate blood supply. Yet, many readers will know someone who has survived a stroke and regained much of their cognitive ability afterwards.

One of the most common problems after stroke is aphasia or language impairment. In a study of 229 adults with stroke, at least 70% of patients showed a reduction in the speed of information processing, and 40% of patients had difficulty with language, memory, visual and spatial tasks, or simple arithmetic [3]. Aphasia is a problem for roughly 38% of patients after stroke, though at least partial recovery of language function occurs spontaneously in some patients within the first 6 months after stroke [4].

Language remediation following stroke is possible, but few treatments are effective, and few patients receive effective treatment. Aphasia treatment can use any of several approaches; it can focus on the language deficit itself, or on strategies designed to compensate for lost language, or on how to use residual skills in communication [5]. Recent interest has centered on the first of these approaches – cognitive linguistic treatment – which aims to facilitate mental processing during speech. This can include helping the patient to hone simple skills such as word-finding, with the assumption that training in fundamental language skills will improve verbal ability.

One of the most crucial variables in language remediation is the intensity of intervention; most patients receive just a few hours of treatment per week and show no more improvement than patients who get no treatment [6]. In contrast, patients who receive aggressive treatment – several hours of help every day – are quite likely to show at least some improvement in language skills by 3 months. In fact, roughly 79% of patients who get aggressive cognitive rehabilitation after stroke show an objective improvement compared to alternative treatments [6]. Thus, high-intensity intervention for aphasia can be an effective way to improve language skills.

Many different types of language remediation have been tried. Not surprisingly, the earliest efforts – which were often of low intensity and sometimes of short duration – tended to result in little or no improvement [7]. Later efforts were more successful, especially if therapies were intensive and focused specifically on those skills with which patients were having the greatest difficulty [8]. If patients get 30–35 h of intervention in just 10 days – an approach called massed practice – there can be a significant improvement in language use within a month. Brain recovery is in some sense analogous to muscle recovery; physical rehabilitation after paralysis is tailored so that the patient is forced to use the paretic hand [9], and language remediation should be tailored so that massed practice is constrained to focus on skills that a patient has lost [4].

One recent study contrasted patients who received constraint-induced aphasia therapy with patients who got a more traditional one-size-fits-all training [4]. Constraint-induced therapy prevents the patient from using compensatory modes of communication, instead constraining the patient to use verbal skills, no matter how impaired. In contrast, the traditional therapy allows – or even encourages – communication by other modes. Constraint-induced therapy also uses language tasks of increasing difficulty in a group setting, where patients are encouraged to play language-based games. Both constraint-induced and traditional treatments were given to patients for 30 h over 10 days, so this was an intensive and exhausting regimen. Language improved with both treatments, and individual analysis showed that 85% of patients benefitted overall. Yet constraint-induced therapy was more effective in the long run, especially if family members also encouraged the patient to rely solely on enhancing verbal skills. Only patients who were actively encouraged by their family to speak were significantly more able to talk 6 months after a stroke. This study was weakened somewhat by the fact that there was no untreated group; because patients tend to improve even in the absence of therapy, it is hard to know if either treatment was better than the healing effects of time.

A great many studies are flawed by the absence of a true control group – a group that does not receive effective treatment. Because most scientists agree that intervention can improve the quality of life for patients, it is unethical to withhold treatment from a needy patient, simply to test if a new intervention is working. But, in the absence of an untreated control group, we cannot be sure if real progress has been made. There are few legitimate ways to compensate for the absence of an untreated control group in stroke research. The most common approach is to contrast two treatments, both of which are thought to work, to determine which treatment is better. But this approach can fail for two reasons. For one thing, patients generally improve with no medical treatment, so an effective intervention has to be very good to be recognized as such. For another thing, in contrasting two treatments – both of which may be somewhat effective – an enormous number of patients must be involved to determine which treatment is truly better.

For example, a recent clinical trial was unable to discern which of two treatments was better, even though 58 stroke patients were enrolled [5]. After stroke, some patients have semantic deficits – the inability to name things or to understand the meaning of words. But patients can also have phonological deficits – the inability to articulate words properly – and semantic and phonological deficits both interfere with speech production. Stroke patients were randomized to treatment either for phonological deficits alone or for both semantic and phonological deficits [5]. At the end of the study, both patient groups were better off than before treatment, but there was no difference between the treatments. Because no untreated patients had been evaluated, we cannot be certain that either treatment was actually better than no treatment at all. Nevertheless, for ethical reasons, it may no longer be possible to do a true placebo-controlled (RBP) trial.

The problem of how to recognize an effective treatment in stroke patients ultimately may force scientists to take very different experimental approaches. One possible approach is to do more research in animals, though animals cannot really provide a satisfying model for human speech. Instead, it may be necessary to study recovery after stroke in the motor region of the brain, despite the reservation that motor recovery in animals may be fundamentally different from language recovery in humans. Nevertheless, research in animals has clearly shown that recovery from a stroke in the motor region requires the brain to literally reorganize and rewire itself [10]. This, of course, suggests that language recovery in humans also happens as the brain reorganizes and rewires itself. This exciting possibility has opened up a second approach that will probably be more useful in humans in the long run; characterizing the process of brain recovery in individual patients as they recover from stroke, using sensitive new methods of analysis. The study of individual patients has shown that language learning after brain injury may indeed be associated with a process of brain rewiring. For example, in children who suffer brain injury during infancy, the language center can develop on the right rather than the left side of the brain [11]. This unusual lateralization of language takes a long time to develop, but such children eventually have language skills that are comparable to normal.

Brain imaging in adult stroke survivors provides evidence that stroke recovery is not simply a result of adapting to a cognitive deficit; it is rather a recovery of lost function [12]. Successful language remediation in adults results in an increase in speed of word-finding [13]. Functional MRI shows that patients with stroke and aphasia initially have a reduced amount of blood flow to that part of the brain that controls speech, but blood flow increases as patients recover their ability to speak. Brain activity elicited by hearing the spoken word generally increases with time during intensive language therapy, as patients improve their ability to discern between words and "pseudowords" [14]. This means that access to word memory improves for patients, so that there is eventually a near-normalization of brain activity during speech [15]. Normalization probably does not mean that all neurons in the injured area have fully recovered; rather, it seems likely that many or most of these neurons have died. But new neurons have been born and older neurons from outside the stroked area have been recruited, to reconstitute language function that had been lost, although the recovered function may reside in a different part of the brain than the original stroked area. Such changes in neuronal role and function are evidence of brain plasticity.

## Pharmacologic Treatment of Aphasia in Stroke

The idea that medication can foster stroke recovery is not new [16]. Pharmacologic treatment for aphasia began as early as 1934, though the results were disappointing. Yet, in recent years, excitement has grown around the idea that certain drugs enhance brain plasticity after stroke [17].

The most successful pharmacotherapy for stroke aphasia uses amphetamine, a psychomotor stimulant [18]. Many forms of amphetamine are available and in wide use, especially as drugs of abuse [e.g., Benzedrine, trimethoxyamphetamine (TMA), 4-methamphetamine (Serenity, Peace, Tranquility), methoxyamphetamine (STP), methylenedioxyamphetamine (MDA), paramethoxy-amphetamine (PMA), and khat]. Controversy surrounds the legitimate use of methlyphenidate (Ritalin), an amphetamine derivative, used in treatment of attention-deficit disorder.

A randomized, double-blind, placebo-controlled (RBP) trial of dextroamphetamine given to patients with aphasia soon after a stroke has shown a clear-cut benefit, compared to placebo. Patients with recent stroke were recruited, tested for language ability, and randomized to receive either amphetamine or an inactive pill that looked just like amphetamine. During the drug trial, every patient received the same speech and language therapy, so the question was whether medication enhanced the effect of language therapy. Each patient received a 1 h session of individualized language therapy every day, with the difficulty of the intervention determined by the patient's performance on a widely used test of language (the Porch Index of Communicative Abilities or PICA). The PICA is a reliable and sensitive measure of change in language use over time. After 5 weeks of training, the language intervention ended, but patients were tested again using the PICA at 6 weeks and 6 months.

Before treatment, there were no differences between patients who got amphetamine and patients who got placebo, because patients had been randomized to treatment [18]. But after 6 weeks, the amphetamine-treated patients had learned more during the training sessions. Even at 6 months – more than 4 months after the language intervention ended – patients who received amphetamine were better able to perform on the PICA. A 15-point increase on the PICA was accepted as "clinically significant"; by this criterion, 83% of the amphetamine-treated patients showed clinical improvement, whereas just 22% of the placebo-treated patients improved to a similar extent. This improvement with amphetamine treatment was striking because patients received a rather low dosage. There were no apparent harmful side effects of treatment, although heart rate and blood pressure did increase transiently after drug administration. The great weakness of this study was that only 21 patients were involved.

Improvement in language ability after amphetamine treatment is thought to result from an increase in brain plasticity [18]. Actual proof for this claim is lacking, though the evidence is intriguing. In experimental animals, administration of dextroamphetamine results in faster recovery of motor function after experimental stroke. Stroke recovery stimulated by dextroamphetamine is further enhanced by training, as if medication works best in synergy with practice. One can sample brain tissue in experimental animals, so we know that behavioral improvement often correlates with actual microscopic changes in the brain, such as enhanced neuronal sprouting or formation of new nerve synapses. If the pharmacologic effects of dextroamphetamine in experimental animals are blocked, the physical and behavioral changes are also blocked, and the animals recover poorly. Experiments involving brain sampling cannot be done in people, of course, but the findings in lab animals suggest that neuronal sprouting may occur in patients who are treated with dextroamphetamine.

The beneficial effects of dextroamphetamine for language learning have now been confirmed in healthy people [17]. A clever study examined healthy people free of stroke, so the goal was not to help people recover from aphasia; rather, the goal was to help well people learn a new vocabulary of 50 made-up words. This was a proper RBP study, with 40 people randomized to receive either dextroamphetamine or an identical-looking inactive placebo. Subjects then had training sessions that lasted 30 min a day for 5 days, during which they learned associations between known objects and pseudowords used to describe those objects. This kind of associative learning is a good model for recovery from aphasia, because aphasic patients must learn what to them would seem to be fresh associations between familiar objects and novel words. After just 1 training session, amphetamine-treated healthy subjects learned pseudowords with significantly greater accuracy than did untreated subjects, and this superior accuracy was retained 1 month and 1 year later. As word-learning was tested, subjects also underwent tests of physiological arousal, including blood pressure, heart rate, and motor speed. No relationship was found between word-learning and blood pressure or heart rate, suggesting that physiological stimulation is not required for word-learning. But amphetamine use did lead to mood elevation, and patients with improved mood were better able to learn words. Treatment effects lasted for at least a year, which is truly striking. In short, amphetamine use led to greater language plasticity and a richer vocabulary in test subjects. Such results are noteworthy because they were attained after less than 3 h of training, rather than the 30 or more hours of training needed for aphasic patients. The same investigators recently reported similar results in an extended study of 60 healthy men [19].

Dextroamphetamine is well tolerated by patients with stroke and it can be safely administered [20], though the risk of recurrent stroke may be high in people who have already had stroke. In healthy people, amphetamine seems to enhance brain efficiency during working memory tasks [21]. Amphetamines have a therapeutic effect on brain activity and mood perhaps because they counteract the way the brain responds to setbacks and frustrations [22]. Amphetamines can increase attentiveness and decrease impulsivity, with no side effects other than food aversion and weight loss [23]. Clearly, if amphetamines improve attentiveness in aphasic patients, this could account for why treated patients are better able to learn new words.

It is thought provoking to note that though amphetamines have a demonstrated effect on language recovery in people following stroke, they have little or no effect on motor recovery [24]. A combined analysis of 287 patients, all of whom were treated with amphetamines to enhance motor recovery, did not show any significant benefit of treatment. Amphetamines even failed to enhance motor recovery in stroke patients if they were combined with physiotherapy [25]. Among healthy subjects, amphetamines do not enhance performance in a sensory task [26]. Furthermore, amphetamines do not increase awareness in patients with severe stroke [27], do not improve motor ability in stroked patients with mild arm paresis [28], do not enhance recovery from severe hemiparetic stroke compared to physiotherapy alone [29], and do not improve motor function in patients in a stroke rehabilitation unit [30]. These findings show a complete absence of efficacy for amphetamine in recovery from a stroke affecting human motor ability. Therefore, studies that characterize motor recovery in animals [10] may not be a relevant model for aphasia recovery in humans.

Evidence for amphetamine efficacy for aphasia is quite strong [31], but there is reasonably good evidence that other drugs also improve aphasia. For example, levodopa (L-dopa) – which is often used to treat Parkinson's disease – has a positive effect on word learning in healthy people [19] and L-dopa improves learning and word retention in healthy people in a dose-dependent way [32]. It seems likely that the next few years will yield progress in stroke treatment because a major hurdle has been crossed; we now know that medical intervention for aphasia can succeed.

# Why Language Learning is So Hard to Study

The conclusion that recovery from stroke-related aphasia can be improved by medical intervention strongly supports the idea that medical intervention can also have an impact on language learning in children. Yet this idea, as attractive as it is, is just a hypothesis – a tentative but testable explanation that guides further research – not a proven fact. And, as it turns out, research into language learning in children is very hard to do.

A great many practical problems complicate research into language learning. For one thing, genes and the environment both have a powerful impact on a child's ability to recognize the spoken word or read the printed text. But there are also genetic influences that have nothing to do with IQ but that can determine a child's ability to discern the sounds of speech (phonemes) or to recognize words [33]. Decoding language requires a child to respond to auditory cues that change rapidly over time [34]. Hence, language learning requires at least four fundamental skills that are intertwined so thoroughly that they are hard to parse apart: listening, speaking, reading, and writing [35]. Children who do poorly at listening are likely also to have problems with reading and writing. This means that it can be hard to identify which skill deficits are primary and need special attention in any given child.

Furthermore, language delay is linked to a range of cognitive and academic problems [36], and disorders that impair language may even herald psychiatric problems [37]. Children who have problems with learning language at 30 months of age are at greater risk of a psychiatric diagnosis – including autism, Aspergers syndrome, or attention-deficit hyperactivity disorder (ADHD) – at 7 years of age [38]. Among children identified with language impairment at age 5, there are often persistent problems at age 19 [39]. Children struggling in school can have problems with information processing, even if they perform well on standardized achievement tests [40]. It turns out that language impairment is a better predictor of long-term academic difficulties than is speech impairment, perhaps because the latter is a mechanical problem whereas the former can be a cognitive problem.

A child who has difficulty in learning language typically has a host of related problems, and it is hard to know how to help such children. Language delay can even cause or aggravate social problems, so children with language delay at a young age are typically also deficient in social interactions and in communication with adults, including teachers [41]. One study concluded that children with language impairment are less able to resolve playground conflicts successfully [42]. This can result in social isolation or bullying, which would further compound psychosocial vulnerabilities.

Yet specific language impairment (SLI) is rather common, affecting 7% of all children [43]. Such children experience problems in understanding and producing spoken language despite having normal intelligence, normal hearing, and normal opportunities to learn language. The cause of SLI is still poorly known, but recent research shows that many language-impaired children are unable to identify specific sounds, especially if those sounds are heard against an acoustically "noisy" background. Under optimal listening conditions of silence, children with language impairment show very subtle speech perception deficits, but these deficits are magnified when speech occurs against a background of noise. Language-impaired children may thus be inefficient at extracting features of speech from noise. In fact, speech impairment may be a fundamental deficit in the way the brain processes sounds or perceives voices [43].

# Why Clinical Trials for Language Impairment are So Hard to Do

Because speech perception is an inherent property of the brain, it is a fruitful and fascinating subject for scientific study. And because speech impairment is so common, it is surprising that the science behind language impairment has not advanced faster. Relatively few current studies examine how remediation impacts children with language impairment, and many of the existing studies are weak or poorly done. Yet language is such an essential life skill that SLI studies are critically important. Hence, it is important to understand those few RBP trials that focus on language impairment in children. Still, it must be confessed that, were we to be rigorous about evaluating *only* RBP trials, there would be nearly nothing to discuss.

Strong RBP trials are hard to do and expensive, so very few have been done in SLI. Rigorous clinical trials are usually done by pharmaceutical companies seeking approval for a new drug that they hope to bring to market. Because a successful medication for a common condition can earn hundreds of millions of dollars, it is worthwhile for a drug company to spend the tens of millions of dollars that may be necessary to do a proper RBP trial. In contrast, few blinded clinical trials have ever been funded to evaluate an intervention in the lives of children, because there is little or no profit to be made.

Strong RBP trials in language impairment are also rare because it is arguably impossible to blind study participants as to whether or not they received an intervention, if the intervention of interest is an educational program. Blinding study participants is not hard to do when sugar pills can be made to look like medication, but how do you blind study participants as to whether or not they received a training program? What if the intervention is a one-on-one interaction between a teacher and a student, designed to help the student respond to the subtle distinctions of speech? What would a believable placebo be? Would a placebo be a teacher who gave no feedback to the student? Or perhaps a teacher who gave incorrect feedback? Clearly, this would not be viable, since a placebo must closely resemble the real intervention.

There can also be an ethical issue involved in using placebo in a clinical trial, because there is a duty to do no harm. If placebo harms a healthy child, that would be indefensible. And, if you believe that an intervention is truly beneficial, it could also be indefensible to withhold treatment from a child merely to do a proper RBP experiment. Yet, if treatment is not withheld from at least some people, you can never really know if an intervention has the desired effect.

A final issue is that a good clinical trial uses endpoints that are determined before the study ever begins. This is important because preselected endpoints are more likely to be objective and unbiased. Imagine that you are testing a new medication for obstructive pulmonary disease; a useful endpoint might be how far a patient can comfortably walk in 6 min. Each patient is measured before treatment and then again after a predetermined interval of treatment. The 6-min walk distance is objective, easy to measure, clinically relevant, and would likely be a useful measure of change. But now imagine that you did a clinical trial, but never thought to measure the 6-min walk distance beforehand. As time goes by, it might become apparent that patients are improving in their ability to walk without fatigue, but you have no measurement made prior to treatment. Consequently, it would be improper to measure the 6-min walk distance after treatment, since you would not know whether patients had actually improved as a result of treatment. Nevertheless, a great many studies of the impact of treatment on language learning use measures that were not determined beforehand. This is particularly likely to happen in studies that last a long time, since changes that result from treatment might be unanticipated or measures used to characterize such change may not have been available when the trial began.

To be as fair and rigorous as possible, we will limit our attention to short-term clinical trials that identify study endpoints beforehand. Because listening, speaking, reading, and writing are so thoroughly convolved [35], we will consider trials that address any of these main components of language. But, because participants cannot be truly blinded as to whether or not they received a training intervention, many studies of educational intervention are excluded from consideration.

## Clinical Trials of Language Remediation in Children

An important impediment to reading is an awareness of how phonemes – the separate sounds that serve as building blocks of words – are manipulated to build words and how words are broken down into phonemes [44]. Phoneme awareness and letter knowledge are thought to be the most basic skills for reading, and children who have a poor awareness of phonemes typically have a hard time learning to read. Recently, a well-designed and compelling clinical trial was done to evaluate a reading intervention based on phoneme awareness. This intervention was given to beginning readers who were progressing poorly in reading. Selection of pupils for inclusion in the study involved testing 636 children in 16 schools, to identify a group of 77 children who were in the bottom 8% of the class in reading development. Children were excluded from the study if they had severe behavior problems, poor cognitive ability, or poor school attendance. Testing was then used to document each child's reading ability at the start of the trial, before children were randomized to receive either the intervention of interest (training in phonological awareness) or a "placebo" that was simply a quiet time without intervention. The intervention itself was a series of 50 training sessions, each of which lasted 20 min, during which children worked in a small group or one-on-one with a teaching assistant. At the end of 10 weeks, all children were tested again. Then children who had previously been in the "placebo" arm of the trial began to take the same training sessions that the "experimental" group of children had received. At the end of 20 weeks, all children were tested yet again. This is a robust and satisfying study design for several reasons: the intervention itself was carefully designed to target a specific problem; the method of testing was defined beforehand; an untreated control group was available for comparison at first; and the group that was trained for 20 weeks could then be compared to a group that was trained for only 10 weeks.

Children who received training showed significant improvement in letter knowledge, word recognition, phoneme awareness, and basic reading skills [44]. The effect of intervention was fairly substantial during the first 10-week period, but tended to plateau during the second 10-week period, as if the children had already obtained a maximal benefit from training. When a group of children was tested again 11 months after training, benefits were still apparent. In truth, the intervention did not change children into top-of-the-class readers; gains tended to be modest but still significant. Not all children benefited to the same extent from the intervention; some children showed a rather large gain, whereas others seemed resistant to change. It is worth emphasizing that this trial was not blinded, since children certainly knew whether or not they were getting an intervention. Nevertheless, children were randomized to treatment and few other trials to date have shown either this level of rigor or this level of success.

The finding of this clinical trial – that training in phoneme awareness can have a positive effect on communication skills – is similar to the results of several other trials that were not as convincing. For example, one trial concluded that phoneme awareness can be trained in very young speech-impaired children, and that training can increase reading success 3 years later [45]. Yet this study was weakened by the fact that only 19 speech-impaired children were studied. Another study, which enrolled 14 students with dyslexia or reading delay, found that reading problems can be remediated after a series of 24 interventions [46], yet this sample size is also too small to inspire confidence.

An ambitious recent study examined the ability of an intervention to improve literacy in children from deprived backgrounds [47]. A cohort of 99 children received training to enhance awareness of the phoneme building blocks of words, while a second group of 114 similar children did not receive an intervention. The intervention occurred once a week for 9 weeks, and lasted less than 1 h each time. It may have been optimistic to think that an intervention that lasted less than 7 h in total could have an impact, but that assumption was made nonetheless. Each training exercise was designed to increase awareness of syllables, rhymes, or phonemes. At an assessment 2 years later, children who received training performed better in tests of rhyme awareness and spelling, but worse in tests of phoneme segmentation. Frankly, this finding makes little sense and is hard to interpret; often, when the results of an experiment are hard to interpret, this is a sign that the experiment was badly done. Though the large sample size would seem to exclude random variation as a cause of the differences in literacy, follow-up testing was 2 years after an intervention that lasted a total of 7 h. This is either too short a training or too long a delay to follow-up for there to have been a strong expectation of benefit. Furthermore, children were not randomized to receive intervention; they were placed in a study group based on their ability, so the experimental and control groups were not comparable at the beginning of the study. One plausible interpretation of the results is that all benefit of intervention had been lost, and children were simply showing random variation in their abilities.

Another intriguing study used a form of computerized training to help dyslexic children learn to read [48]. Yet the data collected in this study – the rate of sound processing by the brain – are so far removed from the classroom that it becomes hard to interpret study results. A group of 44 children with dyslexia were compared to 51 healthy children, after all children received training sessions meant to help children recognize phonemes. At the start of the intervention, dyslexic children were impaired in their ability to hear sounds and to process phonemes, compared to normal children. After the intervention, dyslexic children showed a significant improvement in sound processing, and this improvement lasted for at least 6 months. Unfortunately, the increase in ability to discriminate sounds did not carry over to an ability to read; though training had the specific result anticipated, it did not generalize to the broader goal that was truly desired. This may mean that reading and hearing are not as tightly convolved as investigators had thought, or it may mean that this intervention was simply not an effective means to improve reading scores. Nevertheless, many children have difficulty accurately hearing the sounds of speech [49], and phoneme training may be able to enhance hearing acuity and improve speech comprehension [50].

The failure of computerized interventions to overcome language impairment has now been confirmed by several other studies. One study, which randomized 77 children to receive one of two computer interventions or no intervention at all, found that neither intervention had a significant effect on receptive or expressive language [51]. A second study suggests that children may actually be resistant to computerized training. Although 24 children randomized to receive a computer intervention took over 1,000 training trials each, there were no differences between the 24 trained children and 9 untrained children [52]. A third study found that an intensive auditory training program improved phoneme awareness, but this did not translate to improved reading skills [53]. Perhaps computer training sessions are too complex, or perhaps computer training is boring or inefficient, or perhaps children were trained to do something that is not germane to enhancing language comprehension. In any case, computerized training had little or no impact on language skills. A problem with most computerized interventions is that they are necessarily one-size-fits-all, while a skillful teacher tailors a lesson plan to each individual student. A second problem is that computer interventions tend to be theory-based rather than experience-based. For example, many interventions that seek to enhance phoneme awareness make an assumption that language problems arise from a child's inability to hear fine phonetic distinctions (e.g., the "f" sound in telephone). Yet, if this is not the underlying problem, then interventions designed to treat this deficit might have no effect on language ability.

However, another study – which tried the same computer-based intervention used by several of the negative studies above – found that behavioral remediation can actually change the way the brain works [54]. This study enrolled 20 dyslexic children and compared them to 12 normal children before and after training, using state-of-the-art brain imaging methods. The intervention lasted for an average of 28 training days (100 min per day, 5 days per week), so this was an intensive intervention. At the end of the training period, the dyslexic children had improved significantly in reading and language comprehension, but normal children had not changed at all. A functional magnetic resonance imaging (fMRI) examination showed that dyslexic children given training were able to normalize brain activity related to language use. It is not clear why the results of this study are so much at odds with some of the other studies. However, there has been confirmation by fMRI that language impairment is associated with abnormally low activity in the speech-related regions of the brain [55–58]. In fact, fMRI suggests that language-impaired children actually process meaningful words in much the same way that healthy children process nonsense words; it is as if words lack meaning for language-impaired children.

The overall quality of the science behind language remediation is not yet strong. Some studies enroll an adequate number of language-impaired students, but fail to enroll untreated controls [59, 60]. Thus, scientists cannot know whether improvements in language are a result of normal maturation, which has a strong impact on language [61]. Other studies have followed a tiny number of students over time [62], so that results are really more anecdotal than scientific. Still other studies report that training can result in changes of brain function without proving that such changes result in an actual change in learning outcome. It is not necessarily important that auditory training can improve neural timing in the brainstem [63], if such a change does not also improve language comprehension. Despite these limitations, consensus is growing that, with appropriate instruction, children impaired in reading or language can become more fluent, especially if intervention occurs at a young age [64].

## Methylphenidate in Children with ADHD

It must be stated that we are not advocating pharmacologic intervention for all children or even for all children with some form of language impairment. Yet methylphenidate (Ritalin) is a medical intervention that can be effective for certain children with language impairment. Of course, there are issues and challenges involved in using methylphenidate for patients [65], just as there are side effects for every medication. The most troubling side effect of an amphetamine such as methylphenidate is that humans may show the same kind of sensitization that has been documented in animals given amphetamine. If healthy adults are given a low dose of methylphenidate, there can be general arousal, with mood elevation and an increase in the rate of motor activities such as eye blinking. If volunteers are then given a second dose of methylphenidate 48 h later, the rate of behavioral activation is increased even more than the first time. And if methylphenidate is given a third time, 96 h after the first time, there is a still-greater level of motor activation. What does this mean, in practical terms, for children given Ritalin for ADHD? Does medication leave such children at risk of seizure? The latest results suggest that methylphenidate does not provoke seizures in healthy children [66]. Yet, prevalence of ADHD among children with epilepsy is three- to five fold higher than normal, which may mean that ADHD is either an effect of epilepsy or of epilepsy treatment. Methylphenidate treatment of children prone to seizure is therefore problematic, since it may further increase the risk of epilepsy. Nevertheless, there is evidence that methylphenidate can help children to overcome cognitive limitations.

A robust test of the effect of Ritalin for language impairment occurred in children treated for a form of leukemia known as acute lymphocytic leukemia (ALL). These children often receive prophylactic treatment of the brain, because cancerous ALL cells can lurk there and cause relapse [67]. As a result of therapy directed to the brain, long-term ALL survivors can suffer cognitive impairment. This creates a terrible double-bind for parents; if a child does not receive prophylactic treatment, there is a risk of brain relapse which can have disastrous consequences, yet if a child does receive prophylaxis, there is a risk of cognitive impairment, which can be quite severe. There has been a recent trend to reduce the aggressiveness of brain treatment, because systemic treatment of ALL is becoming more effective. Yet many children still require therapy directed against cancerous cells in the brain and still suffer cognitive impairment years after surviving ALL [68].

A recent clinical trial enrolled 32 children, most of whom had been treated for ALL and all of whom were having trouble concentrating to do school work [69]. These children had an average full-scale IQ of 84.9 – more than 15 points lower than normal – and were impaired in reading, spelling, and math achievement. Children were assessed with a range of tests, then they were randomized to receive either methylphenidate or placebo, in a well-designed RBP trial. Just 90 min after taking their "blinded" medication, children were tested again on a subset of the tests they had just taken. Clearly, this experimental design cannot assess whether children have learned much – other than perhaps how to take the tests – because no educational intervention was attempted. Yet this is a robust way to determine if methylphenidate has had an impact on the ability of children to concentrate and score well on tests. When 17 children who took methylphenidate were compared to 15 children who took placebo, there was a significant and substantial improvement in the ability of medicated children to focus. This could be a result of methylphenidate helping children to use their native abilities more efficiently, to compensate for a disruption of normal brain function. Such improvements in attention and concentration would be expected to result in improved school performance.

The idea that methylphenidate can enhance school performance in cancer patients was then tested by the same investigators, in an elegantly designed RBP trial [70]. A total of 83 cancer patients were given a battery of cognitive tests to establish their abilities at baseline. Then these patients were randomized to receive either methylphenidate or placebo, in identical-looking tablets. After 3 weeks at home, each patient stopped taking whatever they had been taking before and "crossed-over" to taking the other medication, meaning that every patient experienced both methylphenidate and placebo. In addition, two different doses of methylphenidate were tested, to see which dose was more effective. Interestingly, the endpoints used in this study were parent reports and teacher reports of how well the child did while medicated. Because methylphenidate should have an effect on vigilance and sustained attention, and because children were at school during that part of the day during which methylphenidate was expected to have the greatest effect, teacher reports were especially crucial. And because teachers did not know which child was receiving medication, this is an objective way to assess the impact of methylphenidate.

To make a long story short, methylphenidate had a major impact on teacher ratings of ADHD symptoms, classroom attention, academic competence, and social skills [70]. Parents reported a smaller impact of methylphenidate than did teachers, perhaps because medication was wearing off by the time children got home from school. Nevertheless, parents also saw an improvement in ADHD symptoms and attention as a result of treatment. Just 5% of the patients showed side effects that were dose limiting, with insomnia the most common problem, and side effects were mostly seen at the higher dose. Yet, only three patients were taken off high-dose methylphenidate because of side effects. This study is compelling and noteworthy because it used a practical outcome – attention in class, as assessed by the teacher – to determine the efficacy of treatment. But, because this study was only 6 weeks long, we do not yet know whether such benefits are sustained over time.

As with so many things in science, there is conflicting evidence. A study that enrolled ten children with accidental brain injury found no difference between methylphenidate-treated and placebo-treated patients, in attention, memory, or behavior [71], but it is not clear that credence should be given to such a small study. Another study, which enrolled 32 children with central auditory processing disorder, found no effect of methylphenidate on auditory processing, though there was a beneficial effect on sustained attention [72]. Finally, a clinical trial involving 14 children with head injury found that methylphenidate had a small but significant effect on attention [73]. In overview, methylphenidate seems to result in improved concentration and sustained attention among children with brain injury, though some tests were negative [71].

Among children free of brain injury but with a diagnosis of ADHD, results are somewhat more controversial. An RBP trial which enrolled 321 children was able to show that methylphenidate improves teacher ratings of ADHD symptoms, with the only side effects being headache, anorexia, and insomnia [74]. A single daily dose of methylphenidate is sufficient to control ADHD symptoms in most children, and this dose level is tolerated with few side effects [75]. Another RBP trial, which randomized 132 children with ADHD to methylphenidate or placebo, confirmed that methylphenidate improves teacher ratings of ADHD symptoms with few side effects [76]. Amphetamine salts (Adderall XR) have also been both safe and effective in an RBP trial of adolescents with ADHD [23].

Part of the controversy that surrounds use of methylphenidate for ADHD may arise from the fact that children do not all benefit equally from an equal dose of drug [77]. To obtain remission of ADHD symptoms in some children, it may be necessary to treat them with substantially higher doses of methylphenidate. This complicates a traditional RBP trial because dose levels are typically set beforehand, meaning that some children may receive a dose of methylphenidate that is too low to be effective for them. A recent study managed to overcome this problem by using a creative study design; a group of 75 children with ADHD were enrolled in an open-label study, meaning that both parents and teachers knew that children were receiving medication [78]. Methylphenidate dose levels were adjusted in the open-label trial until each child was receiving a dose judged to be effective for them, as determined by teacher ratings of classroom behavior. Only then were all children switched to a blinded medication that looked the same as what they took before. Half the children got "blinded" methylphenidate and half got "blinded" placebo; in other words, half the children got the same medication at the same dose as before, but half the children were weaned off medication entirely. No one knew which children were being weaned, so this part of the trial was truly blinded. By the end of a 2-week withdrawal period, 62% of the children receiving placebo relapsed and were showing symptoms of ADHD, whereas 17% of the children taking methylphenidate had ADHD symptoms. The odds that this difference could have been achieved by chance alone are less than 1-in-1,000, so this is a very convincing study [78].

And yet controversy remains. One RBP study, which evaluated only 26 children treated with methylphenidate, found that medication improved visual-spatial memory and motor control only – not planning, not attention, and not concentration [79]. Another study found that methylphenidate only benefited children who were diagnosed with the "inattentive type" of ADHD, not the more serious "combined type" of ADHD [80]. Nevertheless, those studies that report negative results are generally small and therefore not as compelling. The general finding seems to be that children with ADHD can benefit from methylphenidate.

But does methylphenidate treatment actually improve language ability in children who have ADHD? Several recent studies seem to bear directly on this crucial point. A 6-week study of 45 adolescents, randomized to receive methylphenidate or placebo, found that methylphenidate results in a significant improvement in written language usage, even at a very low dose [81]. In 31 children given methylphenidate in another RBP trial, medication resulted in a significant increase in a child's ability to name colors and recall words quickly, skills which are ordinarily quite difficult for children with ADHD [82].

Finally, and perhaps most convincingly, an RBP trial of 50 children with ADHD found that methylphenidate could improve the ability of children to recount stories [83]. Children listened to an audiotaped story, then were asked to tell the story back to a teacher and to answer a set of questions. Each child's story was recorded, transcribed, and graded for story elements. Children given methylphenidate were better able to understand the actions taken by the main character and the emotional responses of that character. Conversely, methylphenidate had no effect on story comprehension, and no impact on errors that children made in telling the stories.

What this may mean is that children with ADHD have problems understanding and repeating narratives. Story analysis is uniquely sensitive to a cognitive skill that is hard to pinpoint, but that is important to children, who tend to learn a great deal from hearing and telling stories [83].

There has also been a flurry of recent interest surrounding modafinil, a stimulant drug that has been used to treat narcolepsy, but which can also be used to treat ADHD. One strong RBP clinical trial, which enrolled 190 children and randomized them to receive modafinil or placebo for 7 weeks, reported that children who received modafinil had milder symptoms of ADHD [84] with few side effects [85]. Another study of 220 children given modafinil confirmed that ADHD symptoms were ameliorated, but this was an open-label extension trial that did not compare treated children to children given placebo [86]. In any case, there is now robust evidence that medical treatment can alleviate symptoms of ADHD, and can improve school performance and language learning.

## What does Language Remediation Teach Us about Medical Intervention?

One of the most obvious lessons from this overview of language remediation is that intensive programs are far more likely to succeed than programs that are less ambitious. This should be no surprise, as it is consistent with our own experience; "cramming for a test" means all-night studying the night before, not a half-hour study session a month before. In the human mind, repetition builds familiarity and familiarity forges enduring neuronal connections [2]. It is unreasonable to expect someone to learn something complex after a single exposure, just as it was unreasonable to expect that a learning intervention lasting less than 7 h could have an impact on literacy in children [47].

Broad spectrum programs are more likely to be successful in the long run than are narrowly focused programs. This could be because we do not yet understand the fundamental nature of a problem or because we do not yet know how best to treat a problem whose fundamental nature we may understand. For example, there is new evidence that children may be somewhat resistant to computerized training [52], perhaps because such training is not an efficient means to teach. Yet this evidence only emerged after a great deal of time and effort had already been spent on developing computerized training for children. What this suggests is that a skilled teacher who is able to adapt training methods to an individual child cannot yet be surpassed by a one-size-fits-all computer program.

It should be remembered that we are only just beginning to understand the science that will eventually undergird education, and that few established educational approaches have actually been validated scientifically. For example, there was once a trend to teach children to read using a "word-recognition" strategy, even though this is an ineffective method that was never validated in RBP trials. There are probably many educational interventions that are no more successful or sensible than word-recognition, but these approaches await rigorous testing by scientists.

Another obvious lesson is that remediation in childhood is easier and far more feasible than intervention in adulthood. Anyone who has attempted to relearn something in adulthood that they once knew in childhood will know how much easier it was to learn the first time. Perhaps our brains were less distracted or less full of unrelated prior learning when we were children – or perhaps a child's brain is simply more malleable and efficient – but it is far easier to teach a child than an adult. If anyone doubts this, try to relearn the state capitals that you once knew as a child; such rote memorization can be very difficult for an adult but is nearly effortless for a child.

Learning in childhood sets the stage for a great deal of necessary learning in adulthood, and absence of a firm foundation from childhood can be very hard to remediate. For example, poor language use in childhood may reflect poor hearing, yet if the hearing problem is uncorrected, then later language learning may be impaired. This is why early childhood screening for hearing impairment is so vital; it costs little but can save much. A recent study in England compared school districts in which newborn screening was done to other school districts in which such screening was not done [87]. When evaluated at 7–9 years of age, children who were identified as hearing impaired at birth were costing school districts about $4,500 less per year to educate than were hearing-impaired children not identified at birth. Furthermore, children who receive early intervention are less likely to be impaired in language, in social development, and in academic achievement.

It is not yet known if intervention is more likely to succeed in children who are severely impaired or in children who are less impaired. Yet work with youthful violent offenders suggests that school-based intervention programs are more successful in children at high risk of violence [88]. This implies that interventions to treat language impairment should perhaps be directed to the neediest children, though it could also mean that improvement is simply easier to detect in the neediest children. Yet the need for intervention is greater with more impaired children, so it is important to supply help where it is most needed.

Success with an intervention is probably a function of the type of intervention, rather than the problem. Although there are problems that are currently resistant to treatment, there are probably no problems that will remain forever resistant to treatment. In other words, our interventions are likely to improve, so that even difficult problems will become tractable. This claim may be a leap of faith for some, but it is not long ago that many illnesses were thought to be inexorable and inevitable killers.

Children with a problem who do not receive intervention are unlikely to succeed without help. This may be obvious, but there are subtle ramifications; if a successful intervention is available, it is crucial that children actually complete the intervention. For example, education is an intervention that helps children to overcome ignorance, and it is clear that completing that intervention is important. According to the Census Bureau, for a male living in the United States in 2005, average yearly earnings as a function of education were as follows [89]:

| | |
|---|---|
| Less than a high school graduate | $22,138 |
| High school graduate | $31,683 |
| Some college education | $39,601 |
| Bachelor's degree | $53,693 |
| Graduate or professional degree | $71,918 |

Monitoring the success of programs and interventions is essential, otherwise money will be wasted. Validated and reliable standards should be used to make sure that interventions work, but we cannot focus exclusively on long-term outcomes, because this will mean that feedback cycles are too slow. Instead, we must find viable surrogate measures for success. In other words, we need to figure out which short-term changes predict long-term success.

It is important to note that most interventions will produce small effects, but small effects can be important nonetheless. This is especially true when using short-term surrogate endpoints; a small change in the first few weeks after intervention may portend a large and significant change in later years. Unless people are prepared for the fact that no intervention can be immediately and overwhelmingly successful, it is far too easy for critics to claim that "nothing is working."

We must not be discouraged by past failures, for several reasons. First, maximizing human potential is far too important a goal to give up on. Second, the Flynn effect provides strong and unambiguous evidence that medical intervention can increase IQ. Third, the impact of a medical approach to education is only beginning to be felt; RBP trials are likely to have a huge impact on education in the future. And fourth, there has been a great deal of recent and exciting progress in the neurosciences, and this progress promises to put educational interventions on an entirely new and far stronger scientific footing.

All that we know and all that we are learning is consistent with a simple and very compelling truth that should have a powerful effect on the way that we think about educational interventions: if language can be remediated – as it certainly can – then intelligence can be remediated.

# Chapter 12
# Increasing IQ in the United States

Americans tend to feel smug about the health benefits of living in a developed nation, yet a great many people in the United States are too poor to benefit from the advances in medical care that tend to grab the headlines. If tens of millions of people lack medical insurance, then tens of millions will be denied necessary medical treatment. We evaluated 30 medical problems that potentially impair language ability in the United States (Table 9.1), and we concluded that those conditions together may cost the average American child up to 5 IQ points. The problem most likely to result in impaired verbal ability in the United States is rampant poverty that subsumes – and may aggravate – a great many of the other problems that can result from language impairment.

Poverty in the United States is associated with an increased risk of virtually all of the other risk factors for language impairment. Children in poverty are at an increased risk of low birth weight, attention deficit disorder, childhood smoking, lead exposure, asthma, PTSD, childhood neglect, iron deficiency anemia, PCB exposure, diabetes, air pollution, childhood abuse, mercury pollution, epilepsy, and so on [1]. This is a disquieting consideration, given that poverty is far more prevalent in the United States than we would like to admit.

According to the Census Bureau, nearly 37 million Americans – 12.6% of the population – lived in poverty in 2005. The National Academy of Sciences has faulted the Census estimate, claiming that the poverty rate is actually 14.1% [2], meaning that 41.3 million Americans are poverty stricken. The poverty rate has increased in recent years, as earnings have stagnated, but the cost of living continues to rise. By government standards, the poverty line for a family of two parents and two children is $19,806 and – even by these low standards – one child in five in the United States is poor. And there are areas of the country where the news is much worse; in Louisiana, 30% of children live in poverty. Chronic poverty is such a pervasive and intractable problem that there is a temptation to blame the victims for their own misfortune.

In 1994, a book entitled *The Bell Curve: Intelligence and Class Structure in American Life* emerged from a right-wing think tank called the American Enterprise Institute. This book, still one of the most incendiary books ever written about social

R.G. Steen, *Human Intelligence and Medical Illness*, The Springer Series
on Human Exceptionality, DOI 10.1007/978-1-4419-0092-0_12,
© Springer Science+Business Media, LLC 2009

policy, articulated a debate that continues to divide the country. *The Bell Curve* argued that intelligence is genetically determined, and that IQ has woven the fabric of American life. This profoundly pessimistic book claimed that IQ is essentially inalterable, so that it is impossible to intervene in the lives of disadvantaged children. *The Bell Curve* contended that education is an empty exercise, that "... the story of attempts to raise intelligence is one of high hopes, flamboyant claims, and disappointing results. For the foreseeable future, the problems of low cognitive ability are not going to be solved by outside interventions to make children smarter" (p. 389). The authors further argued that, "the more one knows about the evidence, the harder it is to be optimistic about prospects in the near future for raising the [academic] scores of the people who are most disadvantaged by their low scores" (p. 390). This book fueled a passionate and often divisive debate about the efficacy of programs such as Head Start. And yet, *The Bell Curve* was a political polemic masquerading as science, without any foundation in educational practice or the neurobiology of learning. As we have seen, intelligence is not an immutable thing, and intervention can have a major impact on any child's measured IQ.

## The Head Start Program

The purpose of Head Start, according to the Head Start Act of 1981, is "to promote school readiness by enhancing the social and cognitive development of low-income children through the provision, to low-income children and their families, of health, educational, nutritional, social, and other services that are determined, based on family needs assessments, to be necessary." According to the US Administration for Children & Families, the Head Start and Early Head Start programs served 906,993 children in 2006, at a total cost of $6.8 billion. Perhaps because Head Start involved 1,360,000 volunteers as well as 213,000 paid staff, the cost per child in 2006 was just $7,287. This is a remarkably low cost, given that one child in every eight in Head Start had some form of disability (e.g., mild mental retardation, hearing impairment, visual handicap). Furthermore, the ratio of children to paid staff was four to one in Head Start, much better than in most kindergartens.

The Head Start Act required the Secretary of the Department of Health and Human Services to submit interim reports to Congress on the status of children in Head Start at least once every 2 years, between 1999 and 2003. This set a pattern whereby the argument formalized in *The Bell Curve* re-emerged in the spring – like a vile crocus poking through snows of indifference – while program reauthorization was being argued in Congress.

At the core of the ugly debate surrounding the Head Start program was a simple question that lacked a simple answer. Does Head Start help disadvantaged children to succeed? As we have seen before, the best and clearest answer to this question would be attained by a randomized, blinded, placebo-controlled (RBP) trial. But, in the 40-year history of Head Start, such a trial had never been done. This lack of intellectual rigor on the part of Head Start administrators and their Congressional overseers is hard to fathom; since 1965, more than 20 million children have gone through the program and a total of $99.7 billion in appropriations were dedicated to it [3]. Given the public outcry for accountability, the annual cost of Head Start, and the long duration of the program, one would have expected a thorough evaluation of Head Start long ago. But, for whatever reason, this did not happen until recently.

Early results from an ongoing RBP trial – called the National Head Start Impact Study – have emerged, and they are unequivocal: Head Start helps children [4]. The goals of this continuing study are twofold: to determine the impact of Head Start on school readiness of impoverished children; and to assess the circumstances under which Head Start achieves its greatest impact. To meet these ambitious goals, the study involved 383 different Head Start centers in 23 states; each site was selected for inclusion because it was so successful that it attracted more student applicants than it could actually enroll. After the applicants were screened for eligibility, they were randomly assigned (by lottery) to either be enrolled in Head Start or to be wait-listed for a year. Wait-listed children were free to apply to other Head Start programs (or to other similar programs that might be available in the community), but they were formally considered to be "controls" in any case, as they did not receive the intervention under study. We note that it would have been unethical to prevent control children from receiving whatever benefits their parents could obtain for them; there is a moral responsibility to do no harm to study participants. A total of 4,667 children aged 3 or 4 years were randomized to Head Start or to the control group; because this assignment was random, there were few significant differences between the groups at baseline. However, it is worth noting that all of the children enrolled in this study may have had more engaged parents – parents more motivated than average to obtain benefits for their child – than impoverished children whose parents did not try to enroll them in the study.

A great many measurements were made to characterize each child fully, at baseline and at the various follow-up intervals [4]. Four domains were assessed to characterize the impact of Head Start: in the child, cognitive skills, social-emotional skills, and overall health status were measured; and in the parents, parenting skills were studied. Our focus will be squarely on cognitive skills, even though there were significant and often substantial effects of Head Start in the social-emotional, health, and parenting domains. Yet, changes in these other domains were measured by methods that have not been as well validated and that are more open to influence by expectation. Because our attention has so far been on cognition – and because tests to measure cognition are probably more objective – we will largely limit our attention to that domain.

At the end of less than 1 year of Head Start, each child still in the program was assessed again [4]. Unfortunately, there had been some attrition of enrolled children; only 82% of the children screened at baseline were still in the program 1 year later. Such attrition is less than ideal, but it is common, and this is a better rate of follow-up than many studies attain. Children enrolled in Head Start were compared to wait-listed children who were comparable at the beginning of the study, so we can be sure that any differences between groups at the end of 1 year are due to Head Start alone. Even though control children were free to enroll in other programs, only 60% of the control parents found an alternative, meaning that 40% of control children spent the wait-listed year in the care of their parents. Overall, 89% of the Head Start children actually attended Head Start, whereas just 21% of the control children attended another Head Start program.

Cognitive skills improved significantly after a single year of Head Start [4]. Head Start children were able to outperform their peers on several standardized tests (e.g., the Woodcock-Johnson Letter-Word Identification, the Letter-Naming, and the McCarthy Drawing tests), and these differences were so convincing that there was less than one chance in 100 that the differences could be explained by chance alone. Prereading skills of children in Head Start remained below the national norm, but the achievement gap between children of poverty and children of the middle class was substantially reduced. Compared to children in the control group, children in Head Start were 46% closer to the national norm. In short, Head Start was able to roughly halve the expected deficit in reading readiness, when comparing impoverished children to children who enjoyed the benefits of prosperity. Prewriting skills of Head Start children were also lower than expected nationally, but children enrolled in Head Start were 28% closer to the national norm than their peers who did not receive intervention.

The Head Start intervention was most effective for certain well-defined subgroups of children [4]. Head Start worked best for children whose spoken language at home was English. This should not be surprising; roughly 32% of the children had Spanish as a primary language, and these children tended to benefit less from instruction in English reading and writing skills, perhaps because English language skills were not reinforced at home. Head Start was also more effective for minority children. It is unfortunate that, for many people in the United States, impoverishment is not a function of illness or ill-luck; rather it is a function of systematic racism, so that skin color is often a good surrogate measure of income. Because children of color are often victims of discrimination, they may benefit more from an intervention, as a parched seed benefits from water. Head Start was also most effective for those children whose mothers were not depressed; this could be because depressed mothers are less able to offer their children effective reinforcement at home for the lessons learned at school. Finally, Head Start was most helpful, at least in terms of increasing home-reading behavior, for mothers who were older than age 19 at the birth of their child.

The differences between Head Start and control children were not as large as we would like. At the end of 1 year of intervention, Head Start children were still at a disadvantage, compared to middle-class children [4]. Yet, it is critically important that things be kept in context. Children can be impoverished if parents are poorly educated, and poorly educated parents may be unable to provide adequate resources for their children at home. Children can be impoverished if they are part of a minority group that has been systematically discriminated against for generations, and it would be hopelessly naïve to think that a year of remediation could overcome generations of oppression. Children can be impoverished if parents are immigrants or non-native speakers, and such liabilities play out in complex ways across generations. Children can be impoverished if parents are cognitively impaired, and cognitive impairment is often hereditary. Children can be impoverished if parents are drug addicted or alcoholic or otherwise victim to a distorted sense of need, and it is foolish to think that a few hours of intervention could compensate for years of neglect. Children can be impoverished through sheer bad luck, but bad luck can overwhelm the best efforts of parents, and depressed parents may be unable to care for children in supportive ways. Or children can be impoverished because they are abandoned, and such children need years of help to overcome the loss of support that is assumed as a given in the lives of most children.

In addition to the cognitive changes noted, Head Start was successful in other ways as well [4]. Head Start children were more likely to be reported by parents as learning essential literacy skills. Head Start children had significantly fewer behavior problems and fewer symptoms of hyperactivity by parental report. Head Start significantly improved the overall health of enrolled children, with a particularly large impact on dental care. Head Start also had a positive impact on parents; Head Start-affiliated parents were more likely to read to the child at home, more likely to involve the child in enriching activities, less likely to spank the child, and less likely to spank the child frequently. The bottom line is – in less than a year – Head Start significantly and substantially improved the lives and prospects of disadvantaged children.

Other studies have reported similar findings about Head Start. An intervention implemented in Head Start classrooms significantly increased performance on a standardized vocabulary test [5]. Head Start children who received an intervention targeted at improving their understanding of numbers were able to benefit, and the intervention did not reduce the initial positive feelings about school [6]. A study of Head Start children who failed to do well upon school entry found that a prediction could be made as to which children would likely lag behind their peers [7], so aggressive intervention can be targeted to those children who need it most. A small trial that randomized Head Start children to receive social skills training also reported that, as social skills grew, so did vocabulary skills [8].

According to data from the Administration for Children & Families, about 27% of Head Start program staff members in 2006 were parent to a Head Start child, and more than 890,000 parents volunteer in a local Head Start program [3]. This means that people in the best position to know if Head Start actually helps have endorsed the program and are willing to commit their time and effort to it. A review that pooled data from 71 different studies of Head Start found that Head Start had a fairly large impact on cognitive outcome [9]. Evidence for the efficacy of Head Start was actually stronger than evidence for the efficacy of many widely used health interventions, including coronary artery bypass surgery, chemotherapy for breast cancer, or drug therapy for hypertension. True, this review was published in 1993, when the efficacy of bypass surgery, breast cancer chemotherapy, and hypertensive medication was less than it is at present [9]. But it would be unfair to hold Head Start to a higher standard than that to which we hold other health care interventions, and most health interventions are well accepted when their efficacy is no better than that of Head Start. Unfortunately, an aging and self-involved taxpayer may be more willing to pay for their own intervention than to pay for an intervention that benefits the child of a stranger; enlightened self-interest plays a role in any economic decision. Yet aging taxpayers should bear in mind that young people will soon be paying the costs of Medicaid.

## The Early Head Start Program

In 1995, a federal program called Early Head Start was initiated, specifically for low-income pregnant women and families with infants and toddlers [10]. A key to Early Start is that it places great emphasis on serving families as well as children. Soon after the Early Head Start program began, an RBP trial was undertaken to evaluate program success in 3,001 families from 17 urban and rural sites. The goals of this study were several-fold: to determine whether the findings from smaller studies that had gone before – which generally reported benefits to children – could be replicated with a larger sample size; to test the importance of implementing the program exactly as it was designed; and to determine whether Early Start should target children or parents or both.

Children were randomized to get the Early Start intervention or to be wait-listed after their baseline data had been collected, so that wait-listed children could serve as controls [10]. All children received a battery of tests, including several well-accepted and widely used measures of cognition. At baseline, there was only one significant difference between "patients" and controls, so the effort at randomization was successful in allocating children to treatment without bias. Children were enrolled before they were 1 year of age and were followed until they were at least 3 years of age, so the period of follow-up was more than 2 years.

Early Head Start produced positive impacts on the cognitive and language development of children [10]. Among 879 Early Start children, the Bayley Mental Development Index (MDI) was 91.4, whereas the MDI was 89.9 among 779 control children, a difference large enough to be significant. In addition, Early Head Start children scored significantly higher on the Peabody Picture Vocabulary Test, showed less aggressive behavior, were more able to sustain attention, more able to engage with adults, and more cooperative in play situations. Well-implemented programs had better success in helping children, and success was maximized when both parent and child were involved. These results are especially convincing because they are similar to findings in children who were 3- or 4-years old at enrollment, and because so many children were enrolled. In short, there is now robust evidence from well-designed RBP clinical trials that Head Start and Early Head Start are both effective for children who are at risk of performing poorly in school.

There has, of course, been controversy. Evaluation of data from a family literacy program called Even Start, which enrolled 463 families from 18 different sites, was unable to confirm that there was a significant benefit to enrolled children from this program [11]. Perhaps this should not be surprising, as the Early Head Start study [10], which enrolled more than six times as many children, found effects that were significant but not large. Furthermore, the Even Start study was specifically designed to test an idea that seems exceedingly naïve in retrospect (p. 953):

"Family literacy programs are based on the beliefs that children's early learning is greatly influenced by their parents, that parents must develop and value their own literacy skills in order to support their children's educational success, and that *parents are their children's first and best teachers*." (italics added) [11]

But parents are often not literate, and it is hard to imagine how an illiterate parent could be the "first and best teacher" for a child learning to read. As the authors of the study noted (p. 962):

"Even Start parents scored low when compared with national norms based on the general population. When post-tested, the average Even Start parent scored at the fifth percentile (Grade 5.4) on Letter-Word identification, the second percentile on Passage Comprehension (Grade 3.0), the 14th percentile on Word Attack (Grade 3.8), the first percentile on Reading Vocabulary (Grade 3.3), the second percentile on the WJ-R Reading Comprehension cluster (Grade 3.2), and the eighth percentile on the WJ-R Basic Reading Skills cluster (Grade 4.6)." [11]

Such parents may simply be unable to help their children to learn, no matter how well intentioned they are, so that placing the burden of teaching on parents is misguided (p. 966):

"Half of the parents enter [the program] with less than a ninth-grade education, half of the families have an annual income of less than $12,000, and parents and children score at the very lowest levels on literacy assessments (on average, below the fifth percentile). Thus, the changes required on the part of participating parents and children are much greater than previously realized...." [11]

In any case, contrasting 239 families that completed the intervention with 115 families in the control group, there were no differences that achieved significance. Children who received the intervention became more proficient at skills that help in reading and writing, but so too did the control children, and there was no difference in the rate at which the two groups of children learned new skills. Nor were there significant impacts of the intervention on the literacy skills or parenting behaviors of adults. According to the "beliefs" that guided the study, there should be no expectation of improved cognitive scores for children, since the scores of their parents did not improve [11]. What this study means is that education should not be left up to the parents – especially if the parents are functionally illiterate – a lesson that society surely knew centuries ago, when public schools were first established. But this study also sounds a cautionary note; a study that fails to find an effect is not especially noteworthy. Absence of evidence is not the same as evidence of absence. Yet when policy advocates interpret an absence of evidence as proof that an intervention has failed, this is both ignorant and deeply problematic.

An evaluation of programs for parents of infants and toddlers confirms that programs that do not have an impact on parents are unlikely to have an impact on children [12]. Furthermore, for an intervention to be fairly evaluated, children must be randomly assigned to the intervention or to the control group, since nonrandom assignment has caused problems in so many early clinical trials. It is also important to use well-validated measures to assess the impact of intervention, so that measures have clear relevance to the lives of children. It is crucial to replicate findings, to prove that positive results are not an accident or the result of a subtle bias in the study design. Finally, it is important to figure out how much a program costs, so that rational decisions can be made about how to help children most effectively.

There is an oft-repeated criticism that preschool intervention produces a significant short-term change in kindergarten readiness, but not a durable change in intellectual function. This criticism is misleading at best, since it can be used to argue for measures that would be harmful:

"The history of policy making in parenting and early childhood interventions suggests that acting with insufficient data can be wasteful of resources and hope.... In the United States, we have witnessed large investments in major national initiatives, such as Early Head Start and its predecessor the Comprehensive Child Development Program .... Before large investments are made in [clinical trials] or in programs delivered at the community level, fundamental aspects of interventions need to be pretested, piloted, and demonstrated to have potential." [12] (p. 384)

It is certainly true that the Head Start program was remiss in not incorporating a clinical trial design into the first implementation of the program. But that was a different era, one where RBP trials were not so routinely done even in medicine. It is also entirely appropriate to advocate that money be well spent, that any new intervention be tested in an RBP trial, and that clinical trial data support an intervention before it is accepted as standard practice. However, it is a form of analysis/ paralysis to argue that an intervention must be "demonstrated to have potential" before a clinical trial is done. Imagine that a new treatment for breast cancer was tested in a small pilot study and found to result in a 10% increase in 1-year survival for women with a particularly bad form of the disease. Would we require proof that this treatment for breast cancer also improved 10-year survival before we initiated a large-scale clinical trial? In the first analysis, all we can hope for is a positive change in a surrogate measure, which may or may not predict long-term success. To carp that Head Start has not proven that it helps children to succeed in college is mean-spirited at best, and can easily be construed as racist.

## School Readiness and the ABC Project

In 1972 – an era far more utopian than our own cynical time – an experimental project began with the goal of helping at-risk children to succeed in school. The Abecedarian (ABC) Project enrolled low-income infants when they were just 4 months of age, and provided each child with an individualized educational program until the child was old enough for public school [13]. The ABC Project was far ahead of its time because, even though it was not the norm in educational research then (or now), it was a well-designed and rigorous clinical trial. A total of 86 high-risk children were enrolled, with 45 children randomly allocated to receive the ABC intervention. There was no requirement that study children be members of any particular racial or ethnic minority but, because of the demographics of the community in which the study was done – a prosperous college town in the South, whose wealthier members were white – virtually all of the children in the study were African-American. Eight sets of exams were given to children when they were between 6 and 54 months of age. Comparison of those children who got the ABC intervention to control children showed that IQs of children who got the ABC intervention were 8 to 20 points higher than those of control children (average difference = +13.4 points). This difference was highly significant in a statistical sense, with odds that comparable results could be attained by chance alone being 1 in 10,000. By 54 months of age, 31% of control children were below the normal range of IQ, whereas just 7% of children who received the ABC intervention were below the normal range of IQ.

Enrollment in the ABC Project was limited to children who were desperately in need of help [14]. A point system was used to determine eligibility, with life circumstances scored, so that a child had to have many risk factors in order to qualify for ABC enrollment. For example, if a child's mother and father had both completed tenth grade but gone no farther, if the father abandoned the family, and if the household income was less than $5,000 dollars per year (in 1972), such a child would barely qualify for program eligibility. Although a wide range of life circumstances existed among children in the project, the norm was for children to be born to young, single mothers who had not finished high school, and at least 75% of children lacked a father in the home. In short, children in the ABC Project were gravely at risk of growing up in poverty, and of continuing the cycle of poverty into the next generation.

The curriculum that each child experienced was designed to be as much like a game as was possible [14]. The goal was that each "teaching episode" should fully engage the child, and should involve a teacher who was focused on just one or two infants at a time. The curriculum was designed by Dr. Joe Sparling, an educational psychologist at the Frank Porter Graham Child Development Center (FPG Center) at the University of North Carolina in Chapel Hill. He drew on a wide range of sources, including the ideas of Jean Piaget, a developmental psychologist who was a thought-leader at the time. Careful consideration was given to developmental milestones and how to help each child achieve age-appropriate milestones. Finally, parents were asked what goals were most important to them. Sparling said in an interview that, "Parents are very focused on (the idea that) kids should learn to walk at the end of the first year. They're much more focused on that than that (kids) should babble and say some words." So, while parental wishes were incorporated into the curriculum, primacy was given to the task of helping children to achieve appropriate cognitive milestones.

Teachers were encouraged to think about each child individually, and to tailor a learning plan that was appropriate for each child's particular strengths and weaknesses. Sparling believes that getting teachers to think systematically about each child was one of the keys to success. In 1972, the notion of having a teaching plan that was individualized for healthy children was unusual; it was perhaps common practice among teachers of developmentally disabled children, but not among teachers of children who were developmentally normal. Yet, because of this emphasis on one-on-one teaching, Sparling believes that teachers got to know their students far better than was typical of the time. Sparling said that, "What teachers said, at the end of the year, was that, 'I know these kids like I've never known any group of kids before.' It changes the teachers ...."

The program must have been an intensive experience for each child [14]. Classes were held at the FPG Center, which was open each weekday from 7:30 in the morning until 5:30 at night. The hours of the center were long, in part so that parents could use the center as a daycare, which enabled the parents to search for a full-time job. Every child in the program had at least eight contact hours per day, 5 days a week, 50 weeks a year, for more than 4 years, so this was a far more intensive intervention than most other programs of that era or this (Table 12.1). It is possible that the long hours of intervention are responsible for the success of the program; removing a child from a troubled home for 10 h every day has the dual benefit of getting each child into a supportive environment and getting them out of a potentially problematic one. As Sparling said, 34 years after the fact, "We really did have the idea of trying to improve ... school outcomes; we had no idea it was going to (improve) whole-life outcomes."

In some ways, the ABC Project was less than ideal; in a perfect clinical trial, it is expected that the experimental and control groups will differ only in that the experimental group receives the intervention under study. In the ABC Project, "control" children did not receive the same educational curriculum that the "experimental" children got, but they still got some benefits [15]. Children enrolled in the intervention arm came to the FPG Center, where they received breakfast, lunch, and a snack, and study planners realized that this could potentially provide a nutritional benefit to enrolled children. To eliminate the possibility that nutrition alone might raise test scores, all the control children were given iron-fortified formula for the first 15 months of life. In the modern era of breast-feeding, this may not seem like much of a benefit, but virtually no mothers in the study had planned to breast-feed their child. In addition, control children were given free disposable diapers, as an inducement to participate. Finally, control children had ready access to supportive social work services and low-cost medical care through local clinics. Though services provided to the control children were not as extensive as those provided to study-enrolled children, the availability of such services did mean that the greatest difference between the two groups of children was in the educational curriculum itself.

The way that the ABC Project was designed eliminates one of the most telling criticisms that have been leveled against other educational interventions. If a group of children is selected for a new curriculum based on need, then these children are followed over time until they enter school, to whom should they be compared? Should they be compared to other children in the same grade? Yet this is wrong, because it compares children who got an intervention to children who never needed that intervention. In contrast, the ABC Project compared children who got an intervention to other children who also needed but did not receive the intervention, so the impact of the ABC Project has been easier to assess.

Experimental and control children were compared to one another at many time points after the study began [13]. Cognitive testing was done after children had been in the program for 3, 6, 9, 12, 18, 30, 42, and 54 months. Control children were brought into the FPG Center, where they were tested by the same educational

**Table 12.1** Total contact hours for a range of different early-childhood interventions, compared to the contact hours in a typical public school education

| Program | Site | Age enrolled | Hours per week | Weeks per year | Years in program | Contact hours |
|---|---|---|---|---|---|---|
| Early Head Start (EHS) | National | Usually 2–3-years old | 30 | 40 | 2 | 2,400 |
| Head Start (HS) | National | Median 4-years old | 32.5 | 40 | 2 | 2,600 |
| EHS + HS | National | 2-years old | 31 | 40 | 4 | 4,960 |
| Brookline Early Education Project (BEEP) | Brookline, MA | 2–3-years old | 17 | 40 | 5 | 3,400 |
| Milwaukee Project | Milwaukee, WI | 3–6 months | 35 | 50 | 4 | 7,000 |
| Family Development Project | Syracuse, NY | 6–15 months | 30 | 50 | 4 | 6,000 |
| Child–Parent Center Program | Chicago, IL | 3–4-years old | 20 | 40 | 2 | 1,600 |
| Parent–Child Development Center | Houston, TX | 1-year old | 15 | 20 | 2 | ~600 |
| Parent–Child Development Center | Birmingham, AL | 1-year old | ~25 | 25 | 3 | ~2,000 |
| Parent–Child Development Center | New Orleans, LA | 2-months old | 20 | 25 | 4 | ~2,000 |
| Early Training Project | Murfreesboro, TN | 3–4-years old | 21.5 | 10 | 3 | 645 |
| Perry Preschool Program | Ypsilanti, MI | 3 years | 16.5 | 35 | 2 | 1,155 |
| Abecedarian Program | Chapel Hill, NC | 4 months | 40 | 50 | 4 | 8,000 |
| Typical K-12 Public School | National | 5 years | 35 | 40 | 13 | 18,200 |

Data on Early Head Start from Hamm & Ewen, 2006, Center for Law and Social Policy (CLASP) [28]

Data on Head Start from the Head Start web site

Data on the Brookline Early Education Project (BEEP) from Donovan & Cross, National Academies Press [25]

Data on the Milwaukee, Syracuse, and Chicago projects from Feder & Clay, 2002, Investing in the Future of Our Children [24]

Data on the Parent–Child Development Center projects in Houston, Birmingham, and New Orleans from Johnson, 2006 [29]

Data on Early Training Project, Perry Preschool, and the Abecedarian Project from Anderson, 2005, Social Science Research Network [27]

Data supplemented by Wise et al., 2005, Australian Institute of Family Studies [26], and Behrman et al., 1995, The Future of Children [22]

Contact hours are approximated, since the hours per week in a program can vary as a function of age of children and since references are usually rather vague about the number of contact hours provided in a program psychologists who tested all the other ABC children. Hence, it was not possible to "blind" testers as to which child was a control and which had received the curriculum, so the ABC project is not truly a blinded (RBP) trial. But a real effort was made to assure that all children were tested fairly.

Tests that are used to assess the cognitive ability of young children are routinely faulted because, if a child is tested twice, the test scores can differ substantially. This could mean that the tests are simply not very good. Yet Sparling noted, "The fact that scores bounce around, I think has as much to do with infancy as it does with measurement.... So many little behaviors, they're there one day and they're not another. It's almost as if..... behaviors haven't stabilized. As opposed to an older child, if he can say a word on one day, he can say it the next day. With a little kid, that's not necessarily the case." Dr. Frances Campbell, an educational psychologist who actually assessed many of the children, agrees, "I don't really trust a test score to predict very much before a child is three, but the data seem to say that after that age, [results] are pretty predictive." But Campbell also noted that the goal of the study at these early ages was not one of prediction; scientists were simply trying to compare outcomes in groups of children by the fairest and most rigorous means available.

As study children entered public kindergarten at age 5, the regular battery of cognitive testing continued. Standardized tests were used to measure IQ at age 5, age 6.5, age 8, age 12, age 15, and age 21 years, so the trajectory of cognitive development could be characterized [14]. In addition, various other tests were used, to assess school readiness, academic achievement, and so on. One could argue that the constant testing may have helped even those children who did not receive the educational intervention. Constant practice with tests can help children hone their ability to concentrate, and the interest shown in each child by staff at the FPG Center may have validated the idea that intellectual attainment is important. Yet, even if testing had this effect, control children might perform better than community children not involved in the project, but that would not invalidate the project itself. Because practice effects in control children would tend to obscure differences between controls and children who got the ABC intervention, those test differences that do emerge are likely to be very important.

To make a long story short, results achieved by the ABC Project were astonishing. Children who participated in the early intervention had higher cognitive test scores on virtually every test, from age 2 through 21 years [14]. In the most extreme case, the ABC intervention led to a 16-point increase in IQ in 3 years, even though one would not necessarily expect an intervention to lead to *any* IQ change. Bear in mind, IQ tests are designed to focus on aptitude not achievement; psychologists have striven to develop tests that predict future performance, rather than measuring how much a child already knows. Aptitude tests – like the IQ test – seek to measure abstract thinking and the ability to reason, aspects of "intelligence" that have an impact on the ability to achieve in the future, even if past educational opportunities have been limited. Conversely, achievement tests take stock of the knowledge and skills that have already been attained by a child. This is a crucial distinction; a gifted youngster growing up in deprived circumstances may have a great deal of aptitude, but not much in the way of achievement [16].

Overall, the IQ difference between children who got the intervention and children who did not was very large. At 4 to 7 years after the ABC intervention ended, 87% of the experimental children had a normal IQ (greater than 85), whereas just 56% of controls had a normal IQ [17]. The ABC intervention decreased the proportion of children with mild cognitive impairment (IQ $\leq$ 85) by 71%. Children who got the curriculum scored better than children who did not at every time point and on every test. At age 12, more than 4 years after the program ended, the full-scale IQ of ABC-enrolled children averaged at least five points higher that the control children. The conclusion from this clinical trial is that early educational intervention for impoverished children can have long-lasting benefits, in terms of improved cognitive function. Nothing was done in the ABC program that could not be replicated in preschools today, so this is a very profound message of hope.

Early intervention was more effective than late intervention [18]. Optimal performance was associated with early and intensive intervention, and children enrolled in infancy consistently out-performed children enrolled at age 5. A child who received intervention only after age 5 (the normal age to start school) performed no better than a child who got no intervention at all. Early intervention apparently opens a door to later learning. Overall, five factors had an impact on the IQ of a child; maternal IQ, quality of the home environment, parental attitude, infant responsiveness, and gender of the child.

Surely these conclusions are somewhat anticlimactic; otherwise, why are middle-class parents so intent on playing Mozart for a child in the womb, buying geometric mobiles to hang over a crib, reading to children at an early age, or enrolling children in exclusive play-care facilities? Nevertheless, to prove that what is true for middle-class children is also true for impoverished children is a step forward, when many are arguing against early intervention.

Adults who received the ABC intervention as children were less likely to require special education, and consistently achieved higher scores on tests of reading and mathematics [19]. Reading skills were directly related to the length of intervention that a child received. Children who got no intervention scored an 83 on the Woodcock-Johnson test of reading at age 8, well below average. Children who got a 3-year intervention starting in kindergarten scored 86, while children who got a 4-year intervention starting in infancy got a 92, and children who got a 7-year intervention starting in infancy scored a 96, which is not significantly below average. Similar – though less striking – trends were seen for the Woodcock-Johnson test of mathematical ability. At ages 12 and 15, the same trends persisted, with intervention children out-scoring the control children in every case. At the same time that positive changes were seen in the cognitive ability of children, mothers also benefited. Roughly 92% of teenage mothers who enrolled their child in the ABC program had a job by the time their child reached age 15, whereas only 66% of control mothers had a job. Mothers of children in the ABC intervention were able to get more education and earn a better salary than mothers whose children did not get help. In short, intervention improved the cognitive well-being of children and the financial well-being of families.

The ABC intervention reduced the risk of educational failure. ABC children were less likely to fail a grade, more likely to complete high school, and more likely to attend college, and children who got the intervention at an early age still out-performed their peers at age 21 [15]. The full-scale IQ of those who received intervention was 94 at age 21, whereas the IQ of controls was 89. This difference, though it may seem small, was statistically and clinically significant. Children who did not get the educational intervention were able, to some extent, to close the gap over time, so that they caught up to ABC children. But it is hard to know how the control decrement in IQ performance affected early learning experiences at school; it is possible that control ABC subjects were never as comfortable at school and did not learn to achieve to their full potential. What we know is that roughly 36% of adults who got the ABC intervention attended a 4-year college, whereas just 14% of control adults attended college; this difference is so large that there is less than 1 chance in 100 that it can be explained by chance alone ($p < 0.01$). And there is evidence that, the more limited were family resources, the greater were the benefits that children derived from intervention. The persistence of a measured benefit of intervention, from early childhood to adulthood, is almost unique to the ABC project, and may result from the intensity of intervention (Table 10.1).

Surprisingly, the ABC intervention also rewarded study participants with health benefits at age 21 [20]. People who received the ABC intervention had fewer symptoms of depression than did people who acted as controls. Overall, 26% of those people who got the ABC intervention were depressed at age 21, whereas 37% of controls were depressed at the same age. The difference was particularly striking for people with mild depression; half as many experimental subjects had mild depression as did control subjects. In fact, the negative effects of a poor home environment could be almost entirely overcome by the ABC intervention. This is a compelling finding because the risk of depression is much higher than normal among people who grow to maturity in the grip of poverty.

The ABC Project proves that learning begins at the beginning of life [21]. This should not be a surprise. Yet the way in which early intervention played out in the lives of children could not have been predicted. Adults who got the ABC intervention were 42% less likely to become teenage parents ($p < 0.05$), 54% less likely to use marijuana frequently ($p < 0.05$), and 63% more likely to hold a job that required high-level skills ($p < 0.01$). At age 21, the grade-equivalent in mathematics for those who received the ABC intervention was Grade 9.2, whereas the grade-equivalent for controls was Grade 7.9. Over a lifetime, children who got the intervention are projected to earn about $143,000 more from working than are the controls. The clear message from all of this is that infants given a chance to learn will seize it.

What are the strengths of the ABC Project? A major strength, compared to virtually every other program, is that the ABC children got an intensive intervention from a very early age [21]. Another remarkable strength of the ABC Project is that there was a long period of follow-up, so a great deal of data are available; virtually every other study yet done yielded less data, because fewer children were enrolled, the follow-up was shorter, or both. The ABC Project also was able to keep track of virtually every study participant, so that there has been a minimal loss to follow-up. This also means that there is less chance that those children who remain in the study are different in some systematic way from the children who were lost to follow-up.

Yet there were weaknesses with the ABC Project as well [15]. Though it is wonderful to have data from 86 children, it would be better to have data from ten times that many children. Furthermore, every test ever used to assess aptitude or achievement is, to some degree, flawed and imprecise. Much time and effort has been devoted to validating the tests, because of the possibility that they are culturally biased and cannot accurately assess the ability of ethnic minorities [16]. But the ABC Project relied on tests that were potentially flawed to determine if the intervention was working, much the way "No Child Left Behind" relies on flawed tests.

Finally, it is not clear if the results achieved in the ABC Project would have been possible if another sample of children had been studied. Perhaps the extreme neediness of these children primed them to succeed, whereas other children with fewer needs might have had a lesser response to the same intervention. Or perhaps the community within which the intervention was done was important; this study was done in a small, relatively affluent college town with a wide range of human services available and a very competitive school system. Had the study been done in a city with a high crime rate, or a community with a poor school system, or a rural setting with fewer services, it is possible that children would not have benefited as much. Nevertheless, we should acknowledge this study for what it is; a well-designed, well-executed study of a reasonable number of children randomized to treatment and followed for more than 21 years, with very thorough data collection along the way. As such, the ABC study is truly revolutionary.

But the ABC Project is not alone in showing a robust and lasting benefit to children from early intervention. Recent reviews have compiled data from a broad range of early childhood interventions [22–29], and the unequivocal conclusion is that children benefit from early help. Sometimes the benefit of a particular intervention may be slight or statistically not significant, which may mean that program was less effective or that the number of children enrolled was simply too small to detect a meaningful difference. Few studies are as well designed, intensive, or effective as was the ABC Project, so this remains a jewel in the crown of intervention. Yet the weight of evidence for a net benefit from early childhood intervention is overwhelming.

## Other Early Childhood Interventions

Many other early childhood interventions have been done, some of which have achieved a degree of fame within the educational community. Perhaps the best known are the Milwaukee Project [30] and the High/Scope Perry Preschool Program [31], but there was also the Chicago Child–Parent Center Program [32], the Brookline Early Education Project [33], the Syracuse University Family Development Project, the Early Training Project, and several early efforts to assess Head Start as an intervention [21]. The problem is that, with few exceptions [32, 33], these studies have not been published in the peer-reviewed scientific literature. Several of the studies have only been described in book-length monographs, which tend to be long and wordy, written for a narrow audience of believers, and free from the filter of peer review. Peer review can be unfriendly – even hostile – and it is often not helpful, but it is essential to validate work through peer review for the wider audience of people who are interested in education. We will not attempt to review studies here that have only been published in monograph form; we believe that peer review sets a bar for educational research that should not be lowered.

An early childhood program in Chicago, which enrolled 989 children in an intervention and 550 children in a control group, identified several remarkable long-term life changes that resulted from intervention [32]. Low-income minority children, all born in 1980, were recruited to attend early childhood programs at 1 of 25 sites in Chicago. Children who received the intervention had an average of 1.6 years of preschool contact, 1 year of half- or full-day kindergarten, and 1.4 years of school-age participation, compared to controls children who got no preschool, 1 year of full-day kindergarten, and 0.7 years of a school-age program. At the end of the program, children who got the intervention had an average of 4.0 years of contact, whereas controls had 0.7 years of contact. This study was flawed by the fact that children were not randomized prior to treatment; instead, the neediest children from certain areas were assigned to the intervention, then they were matched to children from other areas who acted as controls. Matching was done on the basis of age at school entry, family poverty, and neighborhood poverty. Yet the fact that children were not randomized to treatment meant that there were important differences at the start of the trial. For example, children who got the intervention were somewhat more likely to live in a very poor neighborhood, to have parents who completed high school, and to have fewer siblings. How these differences would play out over time is impossible to predict.

Nevertheless, when all study participants were revisited 15 years later, when they were 21years old, there were crucial differences between controls and those who got the intervention [32]. Children who participated in the preschool intervention for 1 or 2 years had a higher rate of high school graduation, more years of completed education, and lower rates of juvenile arrest. High school graduation was attained by 50% of the intervention group but 39% of the controls; juvenile arrest affected 17% of the intervention group but 25% of the controls; arrest for a violent crime happened to 9% of the intervention group but 15% of the controls; grade failure affected 23% of the intervention group but 38% of the controls; and special education was required by 14% of the intervention group but 25% of the controls. It is interesting to note that the effect of preschool participation was greater for boys than for girls, especially in reducing the drop-out rate. Overall, participation in an early intervention program was associated with better educational and social outcomes up to age 15 years later [32].

A recent review of the Brookline Early Education Project (BEEP) confirms that there are durable benefits from early intervention [33]. The BEEP was a community-based program that provided health and developmental services for children and their families for a period from 3 months before birth through kindergarten entry. A total of 120 young adults who had been enrolled in BEEP 25 years earlier were compared to subjects from the same area who were not enrolled, but who were similar in age, ethnicity, neighborhood, and mother's educational level. Participation in BEEP resulted in children doing better in school and completing at least 1 extra year of schooling. Educational advantages for BEEP participants included learning cognitive skills that helped in planning, organizing, and completing school-related tasks. It seems likely that these skills were responsible for other benefits as well; participants in BEEP were less likely to cost taxpayers additional money for health care, special education, and public assistance in later years.

## Can We Intervene to Augment IQ in Disadvantaged Children?

There is now unambiguous evidence that preschool and school-based interventions can help children to achieve in the short- and long-term. True, there have been many failures, but these have been leavened by some major successes. Yet the evidence is equally clear that preschool programs do not assure success, and that many interventions do not produce durable results. The initial benefits of many programs are diluted over time, and most of the school-based programs are probably too timid or too late to benefit children all that much.

In general, more aggressive interventions yield more striking results. Studies in which budget constraints preclude a time-intensive and costly intervention tend to have disappointing or ambiguous results. And certainly, all interventions fall short of what we would like. It seems that expectations have probably been unrealistic. Just as it was naïve to assume that racial discrimination would end with the Civil Rights Act and that race relations would heal in a few years, it was also unrealistic to assume that the educational debilities experienced by the nation's poorest children could be erased by No Child Left Behind. It will require a sustained effort to level the playing field for all children, so that each child can maximize his or her potential. Yet we now know for a certainty something that was, until quite recently, open to question: every child given a chance to learn will seize it.

As Dr. Edward Zigler, cofounder of the Head Start program, wrote in 2001, before some of the more powerful studies were published [34]:

> "If school readiness is seen as the goal, the evidence is convincing that quality preschool programs work. But if life success indices such as high school graduation and self-sufficiency are construed as the goal, the evidence is weak. The reason is obvious but rarely acknowledged by policy makers and others desiring a quick and easy solution to academic failure: a year or 2 of attending preschool is not an inoculation against all past and future developmental risks imposed by living in poverty. Just as 1 year of good nutrition is not expected to make a child healthy for life, it is foolish to assume that any brief intervention will lead to academic success and a good-paying job." (p. 2379).

# Chapter 13
# Increasing IQ and Social Justice

*"'Tis education forms the common mind;*
*As the twig is bent, the tree's inclined."*

—*Alexander Pope, 1734*

The hypothesis that best explains the Flynn effect is the "rising tide" hypothesis: as public health, in general, improves, children become healthier and better able to demonstrate their innate intellectual ability. Improved health leads to improved mental acuity; if a child is not riven with chronic diarrhea, he will be better able to focus, better able to think, and better able to learn. There is a simple truth here: what is good for the body is good for the brain. Physical health and intellectual health are inseparable; if a child is physically stunted, he will be cognitively stunted as well [1]. However, there is a larger truth here: any intervention that is good for children is good for society, since children are society's most precious resource.

## IQ and Social Justice

The malleability of IQ in infancy and early childhood imposes a duty on every adult. If our society is to be borne forward into the future, adults must care now for children who will later carry the burden of our nation's dreams – and the burden of our own aging bodies. We know beyond doubt that interventions to improve physical health can succeed, especially if they are early and aggressive. Interventions that enhance a child's health will generally raise the IQ [2], and this increases a child's chances for life success in measurable and meaningful ways [3–5]. High quality early child care leads to a durable rise in a child's vocabulary and in scores on IQ tests [6]. Conversely, children who suffer privation early in life are more likely to miss days of school, more likely to have behavioral problems that interfere with learning, and less likely to attend college [7].

Intelligence is inter-generational. The way that intelligence is passed from one generation to the next clearly involves genes, but there is an enormous role for the environment [8]. Mothers who are well-educated tend to have children who are

R.G. Steen, *Human Intelligence and Medical Illness*, The Springer Series
on Human Exceptionality, DOI 10.1007/978-1-4419-0092-0_13,
© Springer Science+Business Media, LLC 2009

school-ready and receptive; mothers who are poorly-educated tend to have children who are unprepared for school and who perform badly on cognitive tests [9]. Well-educated mothers are able to prepare their children to learn, while poorly-educated mothers cannot properly prepare their children for school. Health disparities between well- and poorly-educated mothers account for much of the shortfall in school readiness among the poorest children [10].

But the cycle of poverty can be broken by education. Teenage pregnancy is more common among uneducated young women [11], and teenage mothers are more likely to bear children who live in poverty and struggle in school [12]. Childhood poverty, in turn, leads to early motherhood, which produces increased poverty in the next generation [13], and so the wheel turns. Poverty is associated with troubled parenting [14] and low parental expectations for a child's academic success [15]. Too often, children absorb the expectations of their elders and become pessimistic and prone to failure [16]. But the risk of adolescent pregnancy is reduced by interventions that raise maternal IQ [3]. Such interventions can break the cycle of poverty, benefiting families and future generations in ways that will help IQ to increase progressively over time.

Cognitive disadvantages are just as real as physical disadvantages. But they are easier to ignore, easier to forget, easier to rationalize, easier to discount, and easier to blame on the victim. The only way to break the cycle of poverty is through education. Education must become a part of the health care debate if health care reform is to succeed. If our society seeks to improve public health, economic studies suggests that we will get more return by investing in childhood education than we will by investing in anything else, including adult medical care [17].

What we spend our money on says a lot about who we are as individuals. People spend money in ways that reveal personal choices: perhaps a person has the latest car but little money in the bank; perhaps they live frugally while having a huge stock portfolio; perhaps they pinch pennies to send their children to the best colleges; perhaps children are left to fend for themselves while parents enjoy every luxury; perhaps the whole family lives close to the bone while saving to buy a house or start a business. The money that a person spends clearly reveals something about priorities – who that person is and what they strive to become. A personal budget is a moral accounting; it lays out in stark and unequivocal terms the choices that have been made.

Nations also make choices. The problem is that most citizens never give much thought to these choices; people accept that the government is something apart from themselves, with its own momentum, its own ingrained habits of spending, its own constituencies, its own tendencies and trends, its own accepted "wisdom." People may trust the President, they may like their Senator or Representative, they may accept the word of pundits as to what is wise, they may skim the pages of the newspaper and be overwhelmed by the sheer size of the budget. And it is easy to get lost in the numbers, and fail to realize that national budgets also represent moral choices.

The *New York Times* has estimated that it would cost about $35 billion dollars per year to fund half-day preschool for every 3-year old and full-day preschool for every 4-year old in the United States [18]. Universal preschool would help children meet the stated goals of the No Child Left Behind (NCLB) Act, and would augment the underfunded and much maligned, though highly successful, Head Start program. Currently, Head Start serves less than three of every five eligible children in the nation, leaving 2.6 million eligible children and their families without the means to escape the cycle of poverty [19]. It has been estimated that 37% of all children are born into poverty, yet less than one in three of these children has access to Head Start [20]. Head Start is not only effective, it is also cost-effective [21]; Head Start confers benefits to enrolled children in every cognitive domain evaluated, for less than about $9,000 per child in current dollars.

## Why No Child Left Behind Is a Failure

The NCLB Act requires states, districts, and schools to test children annually for achievement, to report test results separately for all racial and ethnic groups, and to show progressive annual improvements in each test category [22]. The goal of all the prescribed testing is to pressure schools to educate faster and more efficiently. The NCLB then imposes sanctions upon schools that fail to meet goals set by law-makers. Yet early reports suggest that little progress is being made towards those goals; some schools have shown improvements that tend to be small and may be unsustained, but achievement gaps remain and those gaps are widening in many schools. Even many proponents of the NCLB Act now believe that the goals set were unrealistic and cannot be achieved in a decade, as the original time line decreed.

Part of the problem is that NCLB supports an intervention that is too little, too late, and hence destined to fail. The underlying assumption of NCLB seems to be that children fail simply and only because schools permit them to fail; if a child is doing poorly in school, it must be the fault of the teacher. Yet we know that many children fail because of medical circumstances far beyond control of the school. Putting pressure on schools cannot be an effective strategy if children are impaired by medical events that occurred even before they went to school. Though failure of an individual child may be predictable on the basis of public health considerations, schools are nevertheless made to be the scapegoat. However, it is not clear that the goals of the NCLB Act are even achievable, since no effort is made to address the underlying problems that impact children long before they reach school.

Another major problem is that the NCLB creates a perverse inducement for students to drop out of school. According to the *New York Times*, the law mandates that states must test student proficiency frequently using standardized tests, yet the law does not require that students must graduate. Schools are required to show steady improvement in test scores, but there is no penalty if a student leaves school before graduation, so this creates an incentive for failing schools to push students out. If low-achieving students drop out, the overall school performance may rise even if remaining students are no better educated than before. Hence, low-scoring students may be prodded to leave school early and later seek an equivalency certificate, such as a General Educational Development (GED) diploma [23].

Because no effort is made to correct the problems that cause some children to fail, NCLB is essentially a covert attack on the current educational system. The long-term effect of NCLB will likely be to widen the minority achievement gap over time, because the punitive effects of the Act are most damaging to those schools that serve our poorest and most vulnerable children.

NCLB is an experiment in social engineering that lacks any foundation in science. This is ironic, because it is exactly the sort of program that Republicans routinely accuse Democrats of funding; expensive, experimental, intrusive, and ineffectual. Furthermore, NCLB creates a host of troubling moral issues that were recognized by Faithful America, a branch of the National Council of Churches [24]. To summarize their position, NCLB is flawed because:

- It blames schools and teachers for problems that are neither of their making nor within their power to change;
- It places new economic demands on states and on school districts without granting new funding;
- It ranks schools according to test scores of ethnic groups, stigmatizing those children that made a school "fail;"
- It takes funding away from troubled schools, thereby damaging local schools in poor neighborhoods;
- It often deals with troubled students by uprooting them and sending them to successful schools, without any evidence that this will solve the problem;
- It requires children in special education to pass tests that were designed for children free of disabilities;
- It stipulates that children take tests in English, often before children learn the language;
- It obscures the role of humanities, music, and the arts by a relentless focus on testing of reading and arithmetic skills.

## Why Charter Schools Are Not the Way Forward

Charter schools are another example of an educational approach that may be more a distraction than a solution. Charter schools are public schools of choice, yet they have been contentious since they were first established. Charter schools are controversial because they are funded – at least in part – with tax dollars, yet they compete with public schools for students. Any entity can submit a charter to run a school and, if the charter is approved, students apply for admission. Charter schools typically receive a waiver from the local public schools in exchange for a promise to raise student achievement levels. Charters are usually given for 3–5 years, and ideally, academic performance is closely monitored. If academic performance lags behind comparable public schools, the charter may be revoked and the school closed.

Charter schools are popular because they create new educational environments and change is often seen as progress. Charter schools have the freedom to set their own goals, make their own curricula, and teach their own courses, and they do so without charging student tuition. Chartering enables a school to tailor programs to the community, to run autonomously from state-controlled public schools, and to increase parental choice. Charter schools may encourage innovative teaching and provide an alternate career path for teachers, while retaining structure and accountability. Charter schools tend to be smaller than public schools, so they may increase administrative attention to problem students. Charter schools may ultimately be more accountable than typical schools, since the charter is potentially renewed contingent upon success, creating pressure on teachers to foster student achievement.

Nevertheless, charter schools have several disadvantages that may overwhelm the advantages. Charter schools generally have an open admission policy, but they use parental motivation and involvement as a filter. Admission requirements usually cannot exclude students for low test scores or lack of prior achievement, but students can be excluded if their parents simply fail to apply for admission. Admission requirements and lack of information can prevent those students who are most in need or most disadvantaged from applying. In addition, though tuition is not charged, there can be hidden costs – such as uniforms or transportation to and from school – that make it difficult for students in poverty to attend. This means that enrolled students often come from well-off households where education is highly valued, whereas students who are not admitted often come from impoverished households and may need academic help far more desperately. Thus, charter schools may skim the most motivated students from public schools, which enriches charter schools at the expense of public schools.

Public charter school funding also diverts money away from the local school district. Because charter schools are an alternative to public schools, they are separate and can be unequal, which may reconstitute the problem addressed by *Brown v. Board of Education*. In addition, charter schools are a business concern, subject to market forces, though many parents might prefer to insulate their children from the marketplace.

Finally, freedom of choice can mean freedom to be mediocre. Not all innovation is an advance; excluding evolution from a biology curriculum is a bold step back into a past century. Schools that emphasize arts at the expense of sciences may turn out pupils unable to compete for the best colleges and the best jobs. Parents are not all qualified to determine what should be in a school curriculum; the availability of multiple charter school options may enable some parents to assure that their children are overly specialized or inadequately educated. Yet it remains a difficult problem how to resolve the conflicting issues of parental choice and equal student access.

## What Should We Do?

Education is a panacea for most of the ills of society [25]. People are the greatest resource a country has, and an educated populace benefits everyone. Human capabilities are built upon a foundation of preexisting skills, so a well-schooled person is better able to learn new skills on the job or in life. For society, there are many benefits of a good education: health care costs decline because informed people are able to care for themselves and practice preventive medicine; drug and alcohol use wane because people have hope for the future; prison populations shrink because people see law-abiding paths to success; welfare payments diminish because people have the skills to respond to the marketplace; industry enjoys greater productivity because an educated work force can learn new job skills; and retired people can have confidence that their retirement is sustainable because the economy will continue to grow. In short, education is an effective treatment for virtually every major problem that faces our nation today. Yet, for education to be the engine of economic growth that it can be, two things are necessary.

First, we must educate everyone; rich, poor, legal, illegal, black, white, or brown, so that no one is a drag on the system. And the place to start is in preschool. The most cost-effective and attainable new program to implement is a demanding pre-school for every 3- and 4-year old child in the country. This can be patterned after Head Start, but with several significant differences [26]: eligibility should be expanded to younger children and children who might not qualify now; no child should be turned away due to a lack of space; and Head Start teachers should be required to have a Bachelor's degree.

Secondly, we must attract the very best and brightest people into the teaching profession. We give a great deal of lip service to this idea, but we fail to pay teachers a salary commensurate with their responsibilities. If teachers are to help children grow to maturity, if they are to compensate for the failings of parents and society, if they are to raise our children for us, we must pay them what they are worth. According to the National Education Association [27], the average teacher's salary was $47,674 in the most recent survey year available (2004–2005). Over the decade from 1995 to 2005, the average salary for public school teachers increased 0.2% in constant dollars, and 28 states had a decline in teacher salaries after adjusting for inflation. According to the American Federation of Teachers, teacher salaries are comparable to the salaries of other professionals (e.g., accountant, chemist, economist, environmental engineer, registered nurse, computer programmer, compensation analyst). Yet we entrust teachers with a more important responsibility than we do economists and compensation analysts. It seems reasonable to propose that teachers be given a 25% raise, phased in rather quickly, if teachers agree to end the practice of tenure, which keeps non-productive teachers on the payroll. Inertia should be no guarantee of a job; burned-out and ineffective teachers must be eased aside, to make room for the motivated and dynamic people likely to be attracted by higher salaries.

How much would it cost to give teachers a 25% raise? The NEA estimates that roughly $372 billion was spent by federal, state, and local government for primary and secondary schools [28]. Assuming that teacher salaries are 60% of the total cost of school funding, a 25% raise would cost about $56 billion.

Merit pay for highly effective teachers is a simple way to attract the best and brightest to the teaching profession [29]. For years, teachers' unions have vehemently opposed merit pay, and the National Education Association – the largest national teachers' union – still believes it to be "inappropriate," perhaps fearing that merit pay is prone to cronyism. But a consensus is building that financial incentives can attract teachers to work in difficult schools or with difficult students, and can draw bright new recruits into teaching, while energizing veteran teachers and helping them to raise student achievement. In Minnesota, an $86 million initiative designed to professionalize teaching and to provide merit pay has drawn support from teachers as well as from politicians across the political spectrum. A challenge will be how best to determine which teachers deserve a merit increase, since cronyism in awarding bonuses could doom the effort. It is also short-sighted to base merit pay solely on how much student performance improves, because this would potentially dissuade teachers from working with learning-disabled students, who may be harder to help but who need help more desperately.

## The Costs and Benefits of Cognitive Remediation

The *New York Times* estimates that the cost of half-day preschool for 3-year olds and full-day preschool for 4-year olds would be about $35 billion dollars per year [18]. But this calculation is flawed, in that it includes costs, but does not incorporate the benefits that would likely accrue to society from a preschool intervention.

Program costs are always hard to estimate; the Bush Administration initially estimated that the total cost of the war in Iraq would be about $50 billion, but this amount was spent in the first 3 months of 2007 [18]. Benefits are even harder to calculate, because many of the benefits are far in the future, and which benefits to include is controversial. Consequently, any cost-benefit analysis must be taken with a shaker of salt. Nevertheless, we believe that it should be possible to estimate the costs and benefits of a preschool program with reasonable accuracy.

Data from the Perry Preschool program have been used to calculate a detailed cost-benefit analysis for preschool intervention [30]. Because the Perry program was a universal preschool, we will use those numbers to approximate what might result from a national program (Table 13.1). The total cost of the program was roughly $7,600 per child for a year (in 1992, dollars discounted at an inflation-adjusted rate of 3%), so the total program cost per child ($12,356) is a weighted average of the cost for years 1 and 2. This cost is offset by real benefits to program participants, in the form of services that they would not otherwise receive, and real benefits to society as a whole, in the form of reduced costs of K-12 education, increased participant earnings (which are taxable), and reduced crime. In addition, program cost is offset by projected benefits, largely in the form of increased earnings by program participants and reduced costs of crime. On balance, *a preschool program is expected to save taxpayers $76,076 per child*. The cost-benefit ratio of the program (the ratio of the net program benefit to the total program cost) is 7.7, which is extremely high. This average rate of return on investment in preschool education exceeds the average rate of return on investment in the stock market over the last 30 years [30].

Is this analysis plausible? James Heckman, a Nobel Prize winner in economics, has spent a great deal of effort analyzing the costs of investing in disadvantaged children [31]. He proposed that economic models of educational interventions must

**Table 13.1** Present value of the Perry Preschool Program. Cost-benefit is calculated per child and is expressed in 1992 dollars, discounted 3%. Costs appear as negative numbers in parentheses. Cost to society as a whole is the sum of costs to program participants and to tax payers (Summary adapted from Barnett [30])

| Cost and benefits | To program participants | To tax payers | To society as a whole |
|---|---|---|---|
| Total program cost | 0 | (−12,356) | (−12,356) |
| Measured benefits | | | |
|   Child care | 738 | 0 | 738 |
|   K-12 education | 0 | 6,872 | 6,872 |
|   Adult education | 0 | 283 | 283 |
|   College education | 0 | (−868) | (−868) |
|   Taxable earnings | 10,270 | 4,228 | 14,498 |
|   Crime | 0 | 49,044 | 49,044 |
|   Welfare | (−2,193) | 2,412 | 219 |
| Total measured benefits | 8,815 | 61,971 | 70,786 |
| Projected benefits | | | |
|   Taxable earnings | 11,215 | 4,618 | 15,833 |
|   Crime | 0 | 21,337 | 21,337 |
|   Welfare | (−460) | 506 | 46 |
| Total projected benefits | 10,755 | 26,461 | 37,216 |
| Total benefits | 19,570 | 88,432 | 108,002 |
| **Net benefits** | **19,570** | **76,076** | **95,646** |

incorporate two characteristics inherent to learning: early learning makes children more receptive to later learning; and early mastery of basic skills makes it easier to learn advanced skills later. Early intervention is, therefore, more productive than late intervention. Heckman calculates that the rate of return on investment is higher for preschool than for school programs, and far higher than for job-training programs. In fact, the opportunity cost of funds (payout per year if the same dollar was invested in financial assets) makes many school-age programs and virtually all job training programs economically unsustainable in the long run, whereas preschool programs give a rich return on investment. Heckman notes that the rate of return on investment in the Perry Preschool program was 15–17%, and he calculated the cost-benefit ratio as 8.7 (i.e., more than the conservative 7.7 ratio calculated above). Although more than 20% of the American workforce is functionally illiterate (compared to about 10% illiteracy in Germany and Switzerland), Heckman argues that it still makes more sense to spend money on children (p. 1902):

> Investing in disadvantaged young children is a rare public policy initiative that promotes fairness and social justice and at the same time promotes productivity in the economy and in society at large. Early interventions targeted toward disadvantaged children have much higher returns than later interventions such as reduced pupil-teacher ratios, public job training, convict rehabilitation programs, tuition subsidies, or expenditures on police. At current levels of resources, society overinvests in remedial skill investments at later ages and underinvests in the early years.

Several states, most notably Georgia and Oklahoma, have recently implemented universal preschool programs [32]. Georgia implemented the nation's first voluntary pre-kindergarten program, which had enrolled 57% of the state's 4-year-olds by 2004. The program is full-day, so parents are able to search for a full-time job. Oklahoma is first in the nation in the proportion of 4-year-olds enrolled in pre-kindergarten classes; in 2007, 70% of all 4-year-olds in the state were participating. One factor that has been crucial is that pre-kindergarten teachers earn as much as teachers of trigonometry, so there is less turnover as good teachers seek higher salaries. The Oklahoma program, which is patterned after the ABC program that began in North Carolina in 1972, has shown a strongly positive effect on test scores in language and cognition.

How successful is the Oklahoma Educare program? Program participation is voluntary, but parents are generally very enthused [33]. Early results suggest that, all other things being equal, children who go through 1 year of preschool perform 52% better on a test of letter recognition, compared to a child who did not enroll in preschool. The state pays $4,000 per 4-year old child, which is not enough to fund a full-day program, but is far less than the per-child cost of the Perry Preschool program. This may represent an economy of scale, since the Oklahoma program is far larger than the Perry Preschool program. Certainly, there has been no effort to cut corners in the Oklahoma program; every lead teacher must have a Bachelor's degree and there must be one teacher for every 10 students. The program has been so successful that the Governor of Oklahoma is seeking to extend benefits to all 3-year olds in the state.

We know that the long-term costs of ignorance are quite high. So much human knowledge has accumulated that children need to learn faster and more efficiently simply to keep up, just as a new player in the National Basketball Association must be stronger, faster, and better than 30 years ago, just to stay in the game. If we fail our youngest children by not helping them to learn, we will lose a huge number of them to low wages, high crime, and inter-generational failure. We simply cannot afford the price of inaction.

## What Difference Can We Make in the United States?

The ABC intervention resulted in a roughly 5-point increase in IQ [34], which seems quite small. Of what real significance is such a change? Data from the National Longitudinal Survey of Youth (NLSY) suggests that a mere 3-point change in the average IQ of a group can produce a major change in the social behavior of that group [8]. The mean IQ of children in the NLSY was about 100, as expected in any large sample of people in the United States. If this group average is reduced by 3 points, by randomly excluding some people who pull the average IQ up to 100, then a number of striking changes occur. As the IQ of the group decreases to 97, the number of women on welfare increases by 13%, the number of men in prison increases by 11%, the number of people below the poverty line increases by 10%, and the number of children born to a single mother increases by 7%. If the opposite is done, raising IQ by just 3 points by excluding people who pull the average down to 100, then a striking reversal of all these trends is seen. As the average IQ increases to 103, the number of women on welfare decreases by 19%, the number of men in prison decreases by 25%, the number of people below the poverty line decreases by 26%, and the number of children born to a single mother decreases by 17%.

Poverty is a correctable problem that impairs the performance of children. Social justice requires us to intervene. We must insure that every child is fit and able to learn by providing them with adequate prenatal and natal care. We must intervene early and aggressively, so that infants do not suffer privation that will impair their ability to succeed later in life; after all, "as the twig is bent, the tree is inclined." We must provide enriching preschool experiences to every child, so that each child can succeed to the best of their ability. We should abandon no child, let no child endure privation, leave no wound to fester. "Leave no child behind" should become a rule rather than a campaign slogan.

Helping children to become all that their potential will enable [35]:

> is not about government raising children. This is about government strengthening the capacity of families and communities to do the job…. This is not about seeking equality in outcomes. This is about striving for equality of opportunities. This is not about liberals versus conservatives. This is about wise [people] who defy ideological labels….

## What Difference Can the "Rising Tide" Hypothesis Make Overall?

If the "rising tide" hypothesis is correct, then several clear perceptions arise which may have important ramifications in the world. First, the pace of IQ increase will eventually slow in those developed nations – including Canada, France, and the United Kingdom – that already have good health care for infants and children. Yet the pace of IQ increase may continue unabated in those countries – including the United States and many Third World countries – which currently have inadequate health care for the very young.

Secondly, the surest way to improve the long-term outlook for a nation is to improve health care for the children of that nation. Foreign aid to developing nations should therefore include a substantial component of investment in public health measures that improve the lot of the very young. Such investment may prove to be one of the most efficient ways to help a poor country climb out of the trap of poverty, and it is virtually guaranteed to generate good will.

Thirdly, there is a growing risk that the chasm between developed and developing nations will continue to widen – and that the United States will be trapped on the wrong side of that chasm – if public health efforts to deliver quality health care to children are allowed to stagnate.

Finally, every parent knows that people are more willing to die – or to kill – for their children than for any other cause. We can, therefore, expect terrorism, genocide, ethnic cleansing, and oppression to continue unabated if these are perceived as effective tools to right wrongs or to bring greater opportunity to the world's poorest children. Conversely, if structural oppression and inequality of opportunity were to disappear, then people would have less reason to lash out at those who may have oppressed them in the past. In short, social justice requires us to help the poorest children become all they have the potential to be. Mere pragmatism confirms that this is the wisest course.

# References

## Chapter 1

1. Steen, R. G. (1996). *DNA & Destiny: Nature and Nurture in Human Behavior*. New York: Plenum. 259 pp.
2. Flynn, J. (1984). The mean IQ of Americans: Massive gains 1932 to 1978. *Psychological Bulletin, 95*, 29–51.
3. Prifitera, A., Weiss, L. G., & Saklofske, D. H. (1998). The WISC-III in context. In A. Prifitera & D. Saklofske (Eds.), *WISC-III Clinical Use and Interpretation: Scientist-Practitioner Perspectives* (pp. 1–38). New York: Academic.
4. Flynn, J. R. (1998). Israeli military IQ tests: Gender differences small; IQ gains large. *Journal of Biosocial Science, 30*, 541–553.
5. Randhawa, B. S. (1980). Change in intelligence and academic skills of grades four and seven pupils over a twenty-year period. *22nd International Congress of Psychology*. Leipzig, East Germany.
6. de Leeuw, J., & Meester, A. C. (1984). Over het intelligence-onderzoek bijde militaire keuringer vanaf 1925 tot heden [Intelligence-as tested at selections for the military service from 1925 to the present]. *Mens en Maatschappij, 59*, 5–26.
7. Rist, T. (1982). *Det Intellektuelle Prestasjonsnivaet i Befolkningen Sett I lys av den Samfunns-Messige Utviklinga [The level of the intellectual performance of the population seen in the light of developments in the community]*. Oslo, Norway: Norwegian Armed Forces Psychology Service.
8. Teasdale, T. W., & Owen, D. R. (2000). Forty-year secular trends in cognitive abilities. *Intelligence, 28*, 115–120.
9. Bouvier, U. (1969). *Evolution des cotes a quelques tests [Evolution of scores from several tests]*. Brussels, Belgium: Belgian Armed Forces, Center for Research into Human Traits.
10. Elley, W. B. (1969). Changes in mental ability in New Zealand school-children. *New Zealand Journal of Educational Studies, 4*, 140–155.
11. Clarke, S. C. T., Nyberg, V., & Worth, W. H. (1978). *Technical report on Edmonton Grade III achievement: 1956–1977 comparisons*. Edmonton, Alberta: University of Alberta.
12. Uttl, B., & Van Alstine, C. L. (2003). Rising verbal intelligence scores: Implications for research and clinical practice. *Psychology and Aging, 18*, 616–621.
13. Vroon, P. A., de Leeuw, J., & Meester, A. C. (1984). Correlations between the intelligence levels of fathers and sons. In J. R. Flynn (Ed.), Utrecht, The Netherlands: Department of Theoretical Psychology and History of Psychology.
14. Colom, R., & Garcia-Lopez, O. (2003). Secular gains in fluid intelligence: Evidence from the culture-fair intelligence test. *Journal of Biosocial Science, 35*, 33–39.
15. Lynn, R., Hampson, S. L., & Mullineux, J. C. (1987). A long-term increase in the fluid intelligence of English children. *Nature, 328*, 797.

16. Daley, T. C., et al. (2003). IQ on the rise: The Flynn effect in rural Kenyan children. *Psychological Science, 14*, 215–219.
17. Fuggle, P. W., et al. (1992). Rising IQ scores in British children: Recent evidence. *Journal of Child Psychology and Psychiatry, 33*, 1241–1247.
18. Girod, M., & Allaume, G. (1976). L'evolution du niveau intellectuel de la population francaise pendent le dernier quart de siecle [The evolution of the intellectual level of the French population during the last quarter century]. *International Review of Applied Psychology, 25*, 121–123.
19. Steen, R. G. (2007). *The Evolving Brain: The Known and the Unknown* (p. 427). New York: Prometheus Books.
20. Gould, S. J. (1981). *The Mismeasurement of Man*. New York: W. W. Norton & Co. 352 pp.

# Chapter 2

1. Steen, R. G., et al. (2005). Cognitive deficits in children with sickle cell disease. *Journal of Child Neurology, 20*, 102–107.
2. Wechsler, D. (1974). *The Wechsler Intelligence Scales for Children-Revised*. San Antonio, TX: The Psychological Corporation.
3. Sattler, J. M. (1992). Wechsler Intelligence Scale for Children-Revised (WISC-R): description. In *Assessment of children* (p. 995). San Diego, CA: Jerome M. Sattler.
4. Sattler, J. M. (1992). Assessment of ethnic minority children. In *Assessment of Children* (pp. 563–596). San Diego, CA: Jerome M. Sattler.
5. Brown, R. P., & Day, E. A. (2006). The difference isn't black and white: Stereotype threat and the race gap on Raven's Advanced Progressive Matrices. *The Journal of Applied Psychology, 91*, 979–985.
6. Lovaglia, M. J., & Lucas, J. W. (1998). Status processes and mental ability test scores. *American Journal of Sociology, 104*, 195–228.
7. Davis-Kean, P. E. (2005). The influence of parent education and family income on child achievement: The indirect role of parental expectations and the home environment. *Journal of Family Psychology, 19*, 294–304.
8. Prifitera, A., & Saklofske, D. H. (1998). *WISC-III; Clinical Use and Interpretation*. San Diego, CA: Academic. 336 pp.
9. Prifitera, A., Saklofske, D. H., & Weiss, L. G. (2008). *WISC-IV; Clinical Assessment and Intervention* (2nd ed.). San Diego, CA: Academic. 576 pp.
10. Blair, C. (2006). How similar are fluid cognition and general intelligence? A developmental neuroscience perspective on fluid cognition as an aspect of human cognitive ability. *The Behavioral and Brain Sciences, 29*, 109–132.
11. Flynn, J. R. (2006). Towards a theory of intelligence beyond *g*. *The Behavioral and Brain Sciences, 29*, 132–133.
12. Raven, J. (1990). *Standard Progressive Matrices Sets*. Oxford: Oxford Psychologists Press.
13. Gould, S. J. (1981). *The Mismeasurement of Man* (p. 352). New York: W. W. Norton & Co.
14. Brody, N. (1997). Intelligence, schooling, and society. *The American Psychologist, 52*, 1046–1050.
15. Butler, S. R., et al. (1985). Seven-year longitudinal study of the early prediction of reading achievement. *Journal of Educational Psychology, 77*, 349–361.
16. Lubinski, D., & Humphreys, L. G. (1997). Incorporating general intelligence into epidemiology and the social sciences. *Intelligence, 24*, 159–202.
17. Grunau, R. E., Whitfield, M. F., & Fay, T. B. (2004). Psychosocial and academic characteristics of extremely low birth weight (≤ 800 g) adolescents who are free of major impairment compared with term-born control subjects. *Pediatrics, 114*, e725–e732.
18. Bartels, M., et al. (2002). Heritability of educational achievement in 12-year olds and the overlap with cognitive ability. *Twin Research, 5*, 544–553.

19. Peng, Y., et al. (2005). Outcome of low birthweight in China: A 16-year longitudinal study. *Acta Paediatrica, 94*, 843–849.
20. Ivanovic, D. M., et al. (2002). Nutritional status, brain development and scholastic achievement of Chilean high-school graduates from high and low intellectual quotient and socioeconomic status. *The British Journal of Nutrition, 87*, 81–92.
21. Detterman, D. K., & Thompson, L. A. (1997). What is so special about special education? *The American Psychologist, 52*, 1082–1090.
22. Brand, C. (1987). Intelligence testing: Bryter still and bryter? *Nature, 328*, 110.
23. Daley, T. C., et al. (2003). IQ on the rise: The Flynn effect in rural Kenyan children. *Psychological Science, 14*, 215–219.
24. Steen, R. G. (1996). *DNA & Destiny: Nature and Nurture in Human Behavior* (p. 295). New York: Plenum.
25. Flynn, J. R. (1987). Massive IQ gains in 14 nations: What IQ tests really measure. *Psychological Bulletin, 101*, 171–191.
26. Carpenter, P. A., Just, M. A., & Shell, P. (1990). What one intelligence test measures: A theoretical account of the processing in the Raven Progressive Matrices Test. *Psychological Review, 97*, 404–431.
27. Duncan, J., & Owen, A. M. (2000). Common regions of the human frontal lobe recruited by diverse cognitive demands. *Trends in Neurosciences, 23*, 475–483.
28. Flynn, J. R. (1998). Israeli military IQ tests: Gender differences small; IQ gains large. *Journal of Biosocial Science, 30*, 541–553.
29. Teasdale, T. W., & Owen, D. R. (2000). Forty-year secular trends in cognitive abilities. *Intelligence, 28*, 115–120.
30. Teasdale, T. W., & Owen, D. R. (1987). National secular trends in intelligence and education: A twenty-year cross-sectional study. *Nature, 325*, 119–121.

# Chapter 3

1. Steen, R. G. (2007). *The Evolving Brain: The Known and the Unknown.* New York: Prometheus Books.
2. Klein, R. G. (2003). Whither the Neanderthals? *Science, 299*, 1525–1527.
3. Green, R. E., et al. (2006). Analysis of one million base pairs of Neanderthal DNA. *Nature, 444*, 330–336.
4. Bocherens, H., et al. (2005). Isotopic evidence for diet and subsistence pattern of the Saint-Cesaire I Neanderthal: Review and use of a multi-source mixing model. *Journal of Human Evolution, 49*, 71–87.
5. Delson, E., & Harvati, K. (2006). Palaeoanthropology: Return of the last Neanderthal. *Nature, 443*, 762–763.
6. Plagnol, V., & Wall, J. D. (2006). Possible ancestral structure in human populations. *PLoS Genetics, 2*, e105.
7. Krings, M., et al. (1997). Neanderthal DNA sequence and the origin of modern humans. *Cell, 90*, 19–30.
8. Bailey, S. E. (2002). A closer look at Neanderthal postcanine dental morphology: The mandibular dentition. *The Anatomical Record, 269*, 148–156.
9. Mellars, P. (2006). A new radiocarbon revolution and the dispersal of modern humans in Eurasia. *Nature, 439*, 931–935.
10. Kristensen, P., & Bjerkedal, T. (2007). Explaining the relation between birth order and intelligence. *Science, 316*, 1717.
11. Lawlor, D. A., et al. (2005). Early life predictors of childhood intelligence: Evidence from the Aberdeen children of the 1950s study. *Journal of Epidemiology and Community Health, 59*, 656–663.

12. Velandia, W., Grandon, G. M., & Page, E. B. (1978). Family size, birth order, and intelligence in a large South American sample. *American Educational Research Journal, 15*, 399–416.
13. Wagner, M. E., Schubert, H. J., & Schubert, D. S. (1985). Family size effects: A review. *The Journal of Genetic Psychology, 146*, 65–78.
14. Downey, D. B. (2001). Number of siblings and intellectual development: The resource dilution explanation. *The American Psychologist, 56*, 497–504.
15. Wichman, A. L., Rodgers, J. L., & MacCallum, R. C. (2006). A multilevel approach to the relationship between birth order and intelligence. *Personality and Social Psychology Bulletin, 32*, 117–127.
16. Rodgers, J. L., et al. (2000). Resolving the debate over birth order, family size, and intelligence. *The American Psychologist, 55*, 599–612.
17. Garrett, L. (1994). *The Coming Plague: Newly Emerging Diseases in a World Out of Balance.* New York: Penguin Books.
18. Stephens, J. C., et al. (1998). Dating the origin of the CCR5-Δ32 AIDS-resistance allele by the coalescence of haplotypes. *American Journal of Human Genetics, 62*, 1507–1515.
19. Duncan, C. J., & Scott, S. (2005). What caused the Black Death? *Postgraduate Medical Journal, 81*, 315–320.
20. Hedrick, P. W., & Verrelli, B. C. (2006). "Ground truth" for selection on CCR5-Delta32. *Trends in Genetics, 22*, 293–296.
21. Cohn, S. K., & Weaver, L. T. (2006). The Black Death and AIDS: CCR5-Δ32 in genetics and history. *The Quarterly Journal of Medicine, 99*, 497–503.
22. Kremeyer, B., Hummel, S., & Hermann, B. (2005). Frequency analysis of the CCR5 Δ32 HIV resistance allele in a medieval plague mass grave. *Anthropologischer Anzeiger, 63*, 13–22.
23. Bersaglieri, T., et al. (2004). Genetic signatures of strong recent positive selection at the lactase gene. *American Journal of Human Genetics, 74*, 1111–1120.
24. Vroon, P.A., de Leeuw, J., & Meester, A.C. (1984). Correlations between the intelligence levels of fathers and sons. In J. R. Flynn (Ed.), Utrecht, Netherlands: Department of Theoretical Psychology and History of Psychology.
25. Flynn, J. R. (1987). Massive IQ gains in 14 nations: What IQ tests really measure. *Psychological Bulletin, 101*, 171–191.
26. Evans, P. D., et al. (2005). Microcephalin, a gene regulating brain size, continues to evolve adaptively in humans. *Science, 309*, 1717–1720.
27. Mekel-Bobrov, N., et al. (2005). Ongoing adaptive evolution of ASPM, a brain size determinant in *Homo sapiens*. *Science, 309*, 1720–1722.
28. Dorus, S., et al. (2004). Accelerated evolution of nervous system genes in the origin of *Homo sapiens*. *Cell, 119*, 1027–1040.

# Chapter 4

1. Steen, R. G., Spence, D., Wu, S., Xiong, X., Kun, L. E., & Merchant, T. E. (2001). Effect of therapeutic ionizing radiation on the human brain. *Annals of Neurology, 50*, 787–795.
2. Kleinerman, R. A. (2006). Cancer risks following diagnostic and therapeutic radiation exposure in children. *Pediatric Radiology, 36*(Suppl), 121–125.
3. Rilling, J. K., & Insel, T. R. (1999). The primate neocortex in comparative perspective using magnetic resonance imaging. *Journal of Human Evolution, 37*, 191–223.
4. Steen, R. G. (2007). *The Evolving Brain: The Known and the Unknown* (p. 437). New York: Prometheus Books.
5. Caviness, V. S., et al. (1996). The developing human brain: a morphometric profile. In R.W. Thatcher, et al. (Eds.), *Developmental Neuroimaging: Mapping the Development of Brain and Behavior* (pp. 3–14). Academic: San Diego.

6. Giedd, J. (1999). Brain development IX: Human brain growth. *The American Journal of Psychiatry, 156*, 4.
7. Castellanos, F. X., et al. (2002). Developmental trajectories of brain volume abnormalities in children and adolescents with attention-deficit hyperactivity disorder. *Journal of American Medical Association, 288*, 1740–1748.
8. Courchesne, E., et al. (2000). Normal brain development and aging: quantitative analysis at *in vivo* MR imaging in healthy volunteers. *Radiology, 216*, 672–682.
9. James, A. C. D., Crow, T. J., Renowden, S., Wardell, A. M. J., Smith, D. M., & Anslow, P. (1999). Is the course of brain development in schizophrenia delayed? Evidence from onsets in adolescence. *Schizophrenia Research, 40*, 1–10.
10. Raz, N., Gunning-Dixon, F., Head, D., Rodrique, K. M., Williamson, A., & Acker, J. D. (2004). Aging, sexual dimorphism, and hemispheric asymmetry of the cerebral cortex: replicability of regional differences in volume. *Neurobiology of Aging, 25*, 377–396.
11. De Bellis, M. D., et al. (2001). Sex differences in brain maturation during childhood and adolescence. *Cerebral Cortex, 11*, 552–557.
12. Rapoport, J. L., Castellanos, F. X., Gogate, N., Janson, K., Kohler, S., & Nelson, P. (2001). Imaging normal and abnormal brain development: new perspectives for child psychiatry. *The Australian and New Zealand Journal of Psychiatry, 35*, 272–281.
13. Pfefferbaum, A., Mathalon, D. H., Sullivan, E. V., Rawles, J. M., Zipursky, R. B., & Lim, K. O. (1994). A quantitative magnetic resonance imaging study of changes in brain morphology from infancy to late adulthood. *Archives of Neurology, 51*, 874–887.
14. Steen, R. G., et al. (2004). Brain T1 in young children with sickle cell disease: evidence of early abnormalities in brain development. *Magnetic Resonance Imaging, 22*, 299–306.
15. Steen, R. G., Ogg, R. J., Reddick, W. E., & Kingsley, P. B. (1997). Age-related changes in the pediatric brain: quantitative magnetic resonance (qMRI) provides evidence of maturational changes during adolescence. *American Journal of Neuroradiology, 18*, 819–828.
16. Bartzokis, G., Beckson, M., Lu, P. H., Nuechterlein, K. H., Edwards, N., & Mintz, J. (2001). Age-related changes in frontal and temporal lobe volumes in men: a magnetic resonance imaging study. *Archives of General Psychiatry, 58*, 461–465.
17. Kinney, H. C., Brody, B. A., Kloman, A. S., & Gilles, F. H. (1988). Sequence of central nervous system myelination in human infancy. II. Patterns of myelination in autopsied infants. *Journal of Neuropathology and Experimental Neurology, 47*, 217–234.
18. Haynes, R. L., et al. (2005). Axonal development in the cerebral white matter of the human fetus and infant. *The Journal of Comparative Neurology, 484*, 156–167.
19. Casey, B. J., Tottenham, N., Liston, C., & Durston, S. (2005). Imaging the developing brain: what have we learned about cognitive development? *Trends in Cognitive Sciences, 9*, 104–110.
20. Dubois, J., Hertz-Pannier, L., Dehaene-Lambertz, G., Cointepas, Y., & Le Bihan, D. (2006). Assessment of the early organization and maturation of infants' cerebral white matter fiber bundles: a feasibility study using quantitative diffusion tensor imaging and tractography. *NeuroImage, 30*, 1121–1132.
21. Beaulieu, C., et al. (2005). Imaging brain connectivity in children with diverse reading ability. *NeuroImage, 25*, 1266–1271.
22. Ge, Y., Grossman, R. I., Babb, J. S., Rabin, M. L., Mannon, L. J., & Kolson, D. L. (2002). Age-related total gray matter and white matter changes in normal adult brain. Part I: Volumetric MR imaging analysis. *American Journal of Neuroradiology, 23*, 1327–1333.
23. Tisserand, D. J., et al. (2002). Regional frontal cortical volumes decrease differentially in aging: an MRI study to compare volumetric approaches and voxel-based morphometry. *NeuroImage, 17*, 657–669.
24. Giedd, J. N., et al. (1999). Brain development during childhood and adolescence: a longitudinal MRI study. *Nature Neuroscience, 2*, 861–863.
25. Steen, R. G., Hamer, R. M., & Lieberman, J. A. (2007). Measuring brain volume: the impact of precision on sample size in magnetic resonance imaging studies. *American Journal of Neuroradiology, 28*, 1119–1125.

26. Shaw, P., et al. (2006). Intellectual ability and cortical development in children and adolescents. *Nature Neuroscience, 440*, 676–679.

27. Gur, R. C. (2005). Brain maturation and its relevance to understanding criminal culpability of juveniles. *Current Psychiatry Reports, 7*, 292–296.

28. Flieller, A. (1999). Comparison of the development of formal thought in adolescent cohorts aged 10 to 15 years (1967–1996 and 1972–1993). *Developmental Psychology, 35*, 1048–1058.

29. Stoops, N. (2003). U.S. Department of Commerce (Ed.), *Educational Attainment in the United States: Population Characteristics*. U S Census Bureau: Washington, DC.

30. Daley, T. C., Whaley, S. E., Sigman, M. D., Espinosa, M. P., & Neumann, C. (2003). IQ on the rise: the Flynn effect in rural Kenyan children. *Psychological Science, 14*, 215–219.

31. Johnson, S. (2005). *Eveything Bad is Good for You*. New York: Penguin Group

32. Pullman, H., Allik, J., & Lynn, R. (2004). The growth of IQ among Estonian schoolchildren from ages 7 to 19. *Journal of Biosocial Science, 36*, 735–740.

33. Slyper, A. H. (2006). The pubertal timing controversy in the USA, and a review of possible causative factors for the advance in timing of onset of puberty. *Clinical Endocrinology, 65*, 1–8.

34. Herman-Giddens, M. E., et al. (1997). Secondary sexual characteristics and menses in young girls seen in office practice: a study from the Pediatric Research in Office Settings Network. *Pediatrics, 99*, 505–512.

35. Kaplowitz, P. B., Slora, E. J., Wasserman, R. C., Pedlow, S. E., & Herman-Giddens, M. E. (2001). Earlier onset of puberty in girls: relation to increased body mass index and race. *Pediatrics, 108*, 347–353.

36. Sun, S. S., et al. (2002). National estimates of the timing of sexual maturation and racial differences among US children. *Pediatrics, 110*, 911–919.

# Chapter 5

1. Trends in Family Composition. (2006). In Senate Committee on Appropriations, Subcommittee on the District of Columbia.

2. Schneider, B., Atteberry, A., Owens, A. (2005). Family structure and children's educational outcomes. In *Family Matters: Family Structure and Child Outcomes*. Center for Marriage and Families at the Institute for American Values: Birmingham, AL.

3. Press release. (2005). *Income stable, poverty rate increases, percentage of Americans without health insurance unchanged*. US Department of Commerce: Washington, DC.

4. Watson, J. E., et al. (1996). Effects of poverty on home environment: an analysis of three-year outcome data for low birth weight premature infants. *Journal of Pediatric Psychology, 21*, 419–431.

5. Wichman, A. L., Rodgers, J. L., & MacCallum, R. C. (2006). A multilevel approach to the relationship between birth order and intelligence. *Personality and Social Psychology Bulletin, 32*, 117–127.

6. Rodgers, J. L., et al. (2000). Resolving the debate over birth order, family size, and intelligence. *The American Psychologist, 55*, 599–612.

7. Lawlor, D. A., et al. (2005). Early life predictors of childhood intelligence: evidence from the Aberdeen children of the 1950s study. *Journal of Epidemiology and Community Health, 59*, 656–663.

8. Sirin, S. R. (2005). Socioeconomic status and academic achievement: A meta-analytic review of research. *Review of Educational Research, 75*, 417–453.

9. Dickens, W. T., & Flynn, J. R. (2001). Heritability estimates versus large environmental effects: the IQ paradox resolved. *Psychological Review, 108*, 346–369.

10. Steen, R. G. (1996). *DNA & Destiny: Nature and Nurture in Human Behavior*. New York: Plenum, pp. 295.

11. McKenna, M. J., & Kristiansen, A. G. (2007). Molecular biology of otosclerosis. *Advances in Otorhinolaryngology, 65*, 68–74.

12. Ogawa, H., et al. (2007). Etiology of severe sensorineural hearing loss in children: independent impact of congenital cytomegalovirus infection and GJB2 mutations. *Journal of Infectious Diseases, 195*, 782–788.

13. Tibussek, D., et al. (2002). Hearing loss in early infancy affects maturation of the auditory pathway. *Developmental Medicine and Child Neurology, 44*, 123–129.

14. Psarommatis, I. M., et al. (2001). Hearing loss in speech-language delayed children. *International Journal of Pediatric Otorhinolaryngology, 58*, 205–210.

15. Ozmert, E. N., et al. (2005). Relationship between physical, environmental and sociodemographic factors and school performance in primary schoolchildren. *Journal of Tropical Pediatrics, 51*, 25–32.

16. Ross, S. A., et al. (2007). GJB2 and GJB6 mutations in children with congenital cytomegalovirus infection. *Pediatric Research, 61*, 687–691.

17. Needleman, H. (2004). Lead poisoning. *Annual Review of Medicine, 55*, 209–222.

18. Steen, R. G., & Campbell, F. A. (2008). The cognitive impact of systemic illness in childhood and adolescence. In L. Weiss, & A. Prifitera (Eds.), *WISC-IV: Clinical Use and Interpretation* (pp. 365–407). New York: Academic

19. Centers for Disease Control and Prevention. (2006). Adult blood lead epidemiology and surveillance – United States, 2003–2004. *MMWR Morbidity and Mortality Weekly Report, 55*, 876–879.

20. Rogan, W. J., et al. (2005). Lead exposure in children: prevention, detection, and management. *Pediatrics, 116*, 1036–1046.

21. Needleman, H. L., et al. (1979). Deficits in psychologic and classroom performance of children with elevated dentine lead levels. *New England Journal of Medicine, 300*, 689–695.

22. Needleman, H. L., et al. (1990). The long-term effects of exposure to low doses of lead in childhood: an 11-year follow-up report. *New England Journal of Medicine, 311*, 83–88.

23. Bellinger, D. C., Stiles, K. M., & Needleman, H. L. (1992). Low-level lead exposure, intelligence and academic achievement: a long-term follow-up study. *Pediatrics, 90*, 855–861.

24. Canfield, R. L., et al. (2003). Intellectual impairment in children with blood lead concentrations below 10 ug per deciliter. *New England Journal of Medicine, 348*, 1517–1526.

25. Counter, S. A., Buchanan, L. H., & Ortega, F. (2005). Neurocognitive impairment in lead-exposed children of Andean lead-glazing workers. *Journal of Occupational and Environmental Medicine, 47*, 306–312.

26. Koller, K., et al. (2004). Recent developments in low-level lead exposure and intellectual impairment in children. *Environmental Health Perspectives, 112*, 987–994.

27. Tellez-Rojo, M. M., et al. (2006). Longitudinal associations between blood lead concentrations lower than 10 µg/dL and neurobehavioral development in environmentally exposed children in Mexico City. *Pediatrics, 118*, e323–e330.

28. Muniz, P. T., et al. (2002). Intestinal parasitic infections in young children in Sao Paulo, Brazil: prevalences, temporal trends and associations with physical growth. *Annals of Tropical Medicine and Parasitology, 96*, 503–512.

29. Prado, M. S., et al. (2005). Asymptomatic giardiasis and growth in young children; a longitudinal study in Salvador, Brazil. *Parasitology, 131*, 51–56.

30. Simsek, Z., Zeyrek, F. Y., & Kurcer, M. A. (2004). Effect of *Giardia* infection on growth and psychomotor development of children aged 0–5 years. *Journal of Tropical Pediatrics, 50*, 90–93.

31. Celiksoz, A., et al. (2005). Effects of giardiasis on school success, weight and height indices of primary school children in Turkey. *Pediatrics International, 47*, 567–571.

32. Berkman, D. S., et al. (2002). Effects of stunting, diarrhoeal disease, and parasitic infection during infancy on cognition in late childhood: a follow-up study. *Lancet, 359*, 564–571.

33. Larsen, C. S. (2003). Animal source foods and human health during evolution. *Journal of Nutrition, 133*, 3893S–3897S.

34. Poinar, H. N., et al. (2001). A molecular analysis of dietary diversity for three archaic Native Americans. *Proceedings of National Academy of Sciences USA, 98*, 4317–4322.

35. Armelagos, G. J., Brown, P. J., & Turner, B. (2005). Evolutionary, historical and political economic perspectives on health and disease. *Social Science and Medicine, 61*, 755–765.
36. Holliday, R. (2005). Evolution of human longevity, population pressure and the origins of warfare. *Biogerontology, 6*, 363–368.
37. Pearce-Duvet, J. M. (2006). The origin of human pathogens: evaluating the role of agriculture and domestic animals in the evolution of human disease. *Biological Reviews of the Cambridge Philosophical Society, 81*, 369–382.
38. Fevre, E. M., et al. (2005). A burgeoning epidemic of sleeping sickness in Uganda. *Lancet, 366*, 745–747.
39. Cleaveland, S., Laurenson, M. K., & Taylor, L. H. (2001). Diseases of humans and their domestic animals: pathogen characteristics, host range, and the risk of emergence. *Philosophical Transactions of the Royal Society of London B, 356*, 991–999.
40. Diamond, J. (2002). Evolution, consequences, and future of plant and animal domestication. *Nature, 418*, 700–706.
41. Scott, R. S., et al. (2005). Dental microwear texture analysis shows within-species diet variability in fossil hominins. *Nature, 436*, 693–695.
42. Eshed, V., Gopher, A., & Hershkovitz, I. (2006). Tooth wear and dental pathology at the advent of agriculture: new evidence from the Levant. *American Journal of Physical Anthropology, 130*, 145–159.
43. Bonfiglioli, B., Brasili, P., & Belcastro, M. G. (2003). Dento-alveolar lesions and nutritional habits of a Roman Imperial age population (1st-4th c. AD): Quadrella (Molise, Italy). *Homo, 54*, 36–56
44. Kerr, N. W. (1998). Dental pain and suffering prior to the advent of modern dentistry. *British Dental Journal, 184*, 397–399.
45. Lewis, M. E., Roberts, C. A., & Manchester, K. (1995). Comparative study of the prevalence of maxillary sinusitis in later Medieval urban and rural populations in northern England. *American Journal of Physical Anthropology, 98*, 497–506.
46. Stiehm, E. R. (2006). Disease versus disease: how one disease may ameliorate another. *Pediatrics, 117*, 184–191.
47. Pastori, C., et al. (2006). Long-lasting CCR5 internalization by antibodies in a subset of long-term nonprogressors: a possible protective effect against disease progression. *Blood, 107*, 4825–33.
48. Steen, R. G. (2007). *The Evolving Brain: The Known and the Unknown*. New York: Prometheus Books, pp. 437.
49. Hunt, S. A. (2006). Taking heart–cardiac transplantation past, present, and future. *New England Journal of Medicine, 355*, 231–235.
50. DiBardino, D. J. (1999). The history and development of cardiac transplantation. *Texas Heart Institute Journal, 26*, 198–205.
51. Clark, D. A., et al. (1973). Cardiac transplantation in man: review of first three years' experience. *American Journal of Medicine, 54*, 563–576.
52. Pui, C.-H., et al. (2005). Risk of adverse events after completion of therapy for childhood acute lymphoblastic leukemia. *Journal of Clinical Oncology, 23*, 7936–7941.
53. Glaser, V. P. (2000). Investigator profile: E. Donnell Thomas, M.D. *Journal of Hematotherapy and Stem Cell Research, 9*, 403–407

# Chapter 6

1. Schijman, E. (2005). Artificial cranial deformation in newborns in the pre-Columbian Andes. *Childs Nervous System, 21*, 945–950.
2. Anton, S., & Weinstein, K. (1999). Artificial cranial deformation and fossil Australians revisited. *Journal of Human Evolution, 36*, 195–209.
3. Morwood, M. J., et al. (2005). Further evidence for small-bodied hominins from the Late Pleistocene of Flores, Indonesia. *Nature, 437*, 1012–1017.

4. Falk, D., et al. (2005). The brain of LB1, *Homo floresiensis*. *Science, 308,* 242–245.
5. Martin, R. D., et al. (2006). Comment on "The brain of LB1, *Homo floresiensis*". *Science, 312,* 999.
6. Brumm, A., et al. (2006). Early stone technology on Flores and its implications for *Homo Floresiensis*. *Nature, 441,* 624–628.
7. Taylor, A. B., & van Schail, C. P. (2006). Variation in brain size and ecology in *Pongo. Journal of Human Evolution, 52*(1), 59–71.
8. Argue, D., et al. (2006). *Homo Floresiensis*: microcephalic, pygmoid, *Australopithecus*, or *Homo*? *Journal of Human Evolution, 51,* 360–374.
9. Martin, R. D., et al. (2006). Flores hominids: new species or microcephalic dwarf? *The Anatomical Record. Part A, Discoveries in Molecular, Cellular, and Evolutionary Biology, 288,* 1123–1145.
10. Rock, W. P., Sabieha, A. M., & Evans, R. I. W. (2006). A cephalometric comparison of skulls from the fourteenth, sixteenth and twentieth centuries. *British Dental Journal, 200,* 33–37.
11. Lindsten, R., Ogaard, B., & Larsson, E. (2002). Dental arch space and permanent tooth size in the mixed dentition of a skeletal sample from the 14th to the 19th centuries and 3 contemporary samples. *American Journal of Orthodontics and Dentofacial Orthopedics, 122,* 48–58.
12. Luther, P. (1993). A cephalometric comparison of medieval skulls with a modern population. *European Journal of Orthodontics, 15,* 315–325.
13. Jonke, E., et al. (2003). A cephalometric comparison of skulls from different time periods-the Bronze Age, the 19th century and the present. *Collegium Antropologicum, 27,* 789–801.
14. Jantz, R. L., & Jantz, L. M. (2000). Secular change in craniofacial morphology. *American Journal of Human Biology, 12,* 327–338.
15. Boas, F. (1928). *Materials for the Study of Inheritance in Man.* New York: Columbia University Press.
16. Boas, F. (1912). Changes in the bodily form of descendants of immigrants. *American Anthropologist, 14,* 530–562.
17. Sparks, C. S., & Jantz, R. L. (2002). A reassessment of human cranial plasticity: Boas revisited. *Proceedings of the National Academy of Sciences of the United States of America, 99,* 14636–14639.
18. Gravlee, C. C., Bernard, H. R., & Leonard, W. R. (2003). Boas's changes in bodily form: the immigrant study, cranial plasticity, and Boas's physical anthropology. *American Anthropologist, 105,* 326–332.
19. Steen, R. G. (1996). *DNA & Destiny: Nature and Nurture in Human Behavior.* New York: Plenum Press. 295 pp.
20. Jantz, L. M., & Jantz, R. L. (1999). Secular change in long bone length and proportion in the United States, 1800–1970. *American Journal of Physical Anthropology, 110,* 57–67.
21. Federico, G. (2003). Heights, calories and welfare: a new perspective on Italian industrialization, 1854–1913. *Economics and Human Biology, 1,* 289–308.
22. Sunder, M., & Woitek, U. (2005). Boom, bust, and the human body: further evidence on the relationship between height and business cycles. *Economics and Human Biology, 3,* 450–466.
23. Budnik, A., Liczbinska, G., & Gumna, I. (2004). Demographic trends and biological status of historic populations from central Poland: the Ostrow Lednicki microregion. *American Journal of Physical Anthropology, 125,* 369–381.
24. Lewis, M. E. (2002). Impact of industrialization: comparative study of child health in four sties from medieval and postmedieval England (A.D. 850–1859). *American Journal of Physical Anthropology, 119,* 211–223.
25. Nagaoka, T., et al. (2006). Paleodemography of a medieval population in Japan: analysis of human remains from the Yuigahama-minami site. *American Journal of Physical Anthropology, 131,* 1–14.
26. Zarina, G. (2006). The main trends in the paleodemography of the 7th–18th century population of Latvia. *Anthropologischer Anzeiger, 64,* 189–202.
27. Stein, A. D., et al. (2004). Intrauterine famine exposure and body proportions at birth: the Dutch Hunger Winter. *International Journal of Epidemiology, 33,* 831–836.

28. Painter, R. C., et al. (2006). Blood pressure response to psychological stressors in adults after prenatal exposure to the Dutch famine. *Journal of Hypertension, 24*, 1771–1778.

29. Painter, R. C., et al. (2006). Early onset of coronary artery disease after prenatal exposure to the Dutch famine. *American Journal of Clinical Nutrition, 84*, 322–327.

30. Roseboom, T. J., et al. (2001). Effects of prenatal exposure to the Dutch famine on adult disease in later life: an overview. *Twin Research, 4*, 293–298.

31. Painter, R. C., et al. (2005). Adult mortality at age 57 after prenatal exposure to the Dutch famine. *European Journal of Epidemiology, 20*, 673–676.

32. Godfrey, K. M., & Barker, D. J. P. (2000). Fetal nutrition and adult disease. *American Journal of Clinical Nutrition, 71*(Suppl), 1344S–1352S.

33. Barker, D. J. P. (2006). Adult consequences of fetal growth restriction. *Clinical Obstetrics and Gynecology, 49*, 270–283.

34. Barker, D. J. P., et al. (2005). Trajectories of growth among children who have coronary events as adults. *The New England Journal of Medicine, 353*, 1802–1809.

35. Kajantie, E., et al. (2005). Size at birth as a predictor of mortality in adulthood: a follow-up of 350,000 person-years. *International Journal of Epidemiology, 34*, 655–663.

36. Forsen, T. J., et al. (1999). Growth *in utero* and during childhood among women who develop coronary heart disease: a longitudinal study. *British Medical Journal, 319*, 1403–1407.

37. Barker, D. J. P., et al. (2000). Growth *in utero* and blood pressure levels in the next generation. *Journal of Hypertension, 18*, 843–846.

38. Barker, D. J. P. (2001). The malnourished baby and infant. *British Medical Bulletin, 60*, 69–88.

39. Jasienska, G. (2009). Low birth weight of contemporary African Americans: an intergenerational effect of slavery? *American Journal of Human Biology, 21*, 16–24.

# Chapter 7

1. Frances, A., et al. (1994). *Diagnostic and Statistical Manual of Mental Disorders (DSM-IV)* (4th ed.). Washington, DC: American Psychiatric Association.

2. Hoge, C. W., et al. (2004). Combat duty in Iraq and Afghanistan, mental health problems, and barriers to care. *New England Journal of Medicine, 351*, 13–22.

3. Orr, S. P., et al. (2003). Physiologic responses to sudden, loud tones in monozygotic twins discordant for combat exposure: Association with posttraumatic stress disorder. *Archives of General Psychiatry, 60*, 283–288.

4. Lindauer, R. J., et al. (2004). Smaller hippocampal volume in Dutch police officers with posttraumatic stress disorder. *Biological Psychiatry, 56*, 356–363.

5. Bremner, J. D., et al. (2003). MRI and PET study of deficits in hippocampal structure and function in women with childhood sexual abuse and posttraumatic stress disorder. *American Journal of Psychiatry, 160*, 924–932.

6. Carrion, V. G., Weems, C. F., & Reiss, A. L. (2007). Stress predicts brain changes in children: A pilot longitudinal study on youth stress, posttraumatic stress disorder, and the hippocampus. *Pediatrics, 119*, 509–516.

7. Tischler, L., et al. (2006). The relationship between hippocampal volume and declarative memory in a population of combat veterans with and without PTSD. *Annals of the New York Academy of Sciences, 1071*, 405–409.

8. Macklin, M. L., et al. (1998). Lower precombat intelligence is a risk factor for posttraumatic stress disorder. *Journal of Consulting and Clinical Psychology, 66*, 323–326.

9. Buckley, T. C., Blanchard, E. B., & Neill, W. T. (2000). Information processing and PTSD: A review of the empirical literature. *Clinical Psychology Review, 20*, 1041–1065.

10. Haier, R. J., et al. (2004). Structural brain variation and general intelligence. *Neuroimage, 23*, 425–433.

11. Saigh, P. A., et al. (2006). The intellectual performance of traumatized children and adolescents with or without posttraumatic stress disorder. *Journal of Abnormal Psychology, 115*, 332–340.
12. Breslau, N., Lucia, V. C., & Alvarado, G. F. (2006). Intelligence and other predisposing factors in exposure to trauma and posttraumatic stress disorder: A follow-up study at age 17 years. *Archives of General Psychiatry, 63*, 1238–1245.
13. Steen, R. G., Hamer, R. M., & Lieberman, J. A. (2007). Measuring brain volume: The impact of precision on sample size in magnetic resonance imaging studies. *American Journal of Neuroradiology, 28*, 1119–1125.
14. De Bellis, M. D., & Kuchibhatla, M. N. (2006). Cerebellar volumes in pediatric maltreatment-related posttraumatic stress disorder. *Biological Psychiatry, 60*, 697–703.
15. Woodward, S. H., et al. (2006). Hippocampal volume, PTSD, and alcoholism in combat veterans. *American Journal of Psychiatry, 163*, 674–681.
16. Samuelson, K. W., et al. (2006). Neuropsychological functioning in posttraumatic stress disorder and alcohol abuse. *Neuropsychology, 20*, 716–726.
17. Delaney-Black, V., et al. (2002). Violence exposure, trauma, and IQ and/or reading deficits among urban children. *Archives of Pediatrics and Adolescent Medicine, 156*, 280–285.
18. Cho, K. (2001). Chronic "jet lag" produces temporal lobe atrophy and spatial cognitive deficits. *Nature Neuroscience, 4*, 567–568.
19. Sokol, D. K., et al. (2003). From swelling to sclerosis: Acute change in mesial hippocampus after prolonged febrile seizures. *Seizure, 12*, 237–240.
20. Frank, G. K., et al. (2004). Neuroimaging studies in eating disorders. *CNS Spectrums, 9*, 539–548.
21. Moore, G. J., et al. (2000). Lithium-induced increase in human brain grey matter. *Lancet, 356*, 1241–1242.
22. Hardmeier, M., et al. (2003). Atrophy is detectable within a 3-month period in untreated patients with active relapsing-remitting multiple sclerosis. *Archives of Neurology, 60*, 1736–1739.
23. Hoogervorst, E. L., Polman, C. H., & Barkhof, F. (2002). Cerebral volume changes in multiple sclerosis patients treated with high-dose intravenous methylprednisolone. *Multiple Sclerosis, 8*, 415–419.
24. Gilbert, A. R., et al. (2000). Decrease in thalamic volumes of pediatric patients with obsessive-compulsive disorder who are taking paroxetine. *Archives of General Psychiatry, 57*, 449–456.
25. Denton, E. R., et al. (2000). The identification of cerebral volume changes in treated growth-hormone deficient adults using serial 3D MR image processing. *Journal of Computer Assisted Tomography, 24*, 139–145.
26. Walters, R. J., et al. (2001). Haemodialysis and cerebral oedema. *Nephron, 87*, 143–147.
27. Lieberman, J. A., et al. (2005). Antipsychotic drug effects on brain morphology in first-episode patients. *Archives of General Psychiatry, 62*, 361–370.
28. Garver, D. L., et al. (2000). Brain and ventricle instability during psychotic episodes of the schizophrenias. *Schizophrenia Research, 44*, 11–23.
29. Pfefferbaum, A., et al. (1995). Longitudinal changes in magnetic resonance imaging of brain volumes in abstinent and relapsed alcoholics. *Alcoholism, Clinical and Experimental Research, 19*, 1177–1191.
30. Pfefferbaum, A., et al. (2004). Brain volumes, RBC status, and hepatic function in alcoholics after 1 and 4 weeks of sobriety: Predictors of outcome. *American Journal of Psychiatry, 161*, 1190–1196.
31. Agartz, I., et al. (2003). MR volumetry during acute alcohol withdrawal and abstinence: A descriptive study. *Alcohol and Alcoholism, 38*, 71–78.
32. Herrnstein, R. J., & Murray, C. (1994). *The Bell Curve: Intelligence and Class Structure in American Life* (p. 845). New York: The Free Press.
33. (2008, January 26) The starvelings; malnutrition (New studies of the impact of hunger on child's health). *The Economist, 386*, 8564.
34. Olness, K. (2003). Effects on brain development leading to cognitive impairment: A worldwide epidemic. *Journal of Developmental and Behavioral Pediatrics, 24*, 120–130.

35. Barbour, V., et al. (2008). Scaling up international food aid: Food delivery alone cannot solve the malnutrition crisis. *PLoS Medicine, 5*, 1525–1527. (e235).

36. Steen, R. G. (2007). *The Evolving Brain: The Known and the Unknown* (p. 437). New York: Prometheus Books.

37. Galler, J. R., & Barrett, L. R. (2001). Children and famine: Long-term impact on development. *Ambulatory Child Health, 7*, 85–95.

38. Galler, J. R., Ramsey, F., & Solimano, G. (1984). The influence of early malnutrition on subsequent behavioral development. III. Learning disabilities as a sequel to malnutrition. *Pediatric Research, 18*, 309–313.

39. Galler, J. R., et al. (1987). Long-term effects of early kwashiorkor compared with marasmus. II. Intellectual performance. *Journal of Pediatric Gastroenterology and Nutrition, 6*, 847–854.

40. Galler, J. R., et al. (1990). The long-term effects of early kwashiorkor compared with marasmus. IV. Performance on the national high school entrance examination. *Pediatric Research, 28*, 235–239.

41. Galler, J. R., et al. (2004). Postpartum maternal moods and infant size predict performance on a national high school entrance examination. *Journal of Child Psychology and Psychiatry, 45*, 1064–1075.

42. Powell, C. A., et al. (1995). Relationships between physical growth, mental development and nutritional supplementation in stunted children: The Jamaica study. *Acta Paediatrica, 84*, 22–29.

43. Mendez, M. A., & Adair, L. S. (1999). Severity and timing of stunting in the first two years of life affect performance on cognitive tests in late childhood. *Journal of Nutrition, 129*, 1555–1562.

44. Ivanovic, D. M., et al. (2002). Nutritional status, brain development and scholastic achievement of Chilean high-school graduates from high and low intellectual quotient and socioeconomic status. *British Journal of Nutrition, 87*, 81–92.

45. Hoddinott, J., et al. (2008). Effect of a nutrition intervention during early childhood on economic productivity in Guatemalan adults. *Lancet, 371*, 411–416.

46. Berkman, D. S., et al. (2002). Effects of stunting, diarrhoeal disease, and parasitic infection during infancy on cognition in late childhood: A follow-up study. *Lancet, 359*, 564–571.

47. Carey, W. B. (1985). Temperament and weight gain in infants. *Journal of Developmental and Behavioral Pediatrics, 6*, 128–131.

48. Vohr, B. R., et al. (2006). Beneficial effects of breast milk in the neonatal intensive care unit on the developmental outcome of extremely low birth weight infants at 18 months of age. *Pediatrics, 118*, e115–e123.

49. Slykerman, R. F., et al. (2005). Breastfeeding and intelligence of preschool children. *Acta Paediatrica, 94*, 832–837.

50. Lawlor, D. A., et al. (2006). Early life predictors of childhood intelligence: Findings from the Mater University study of pregnancy and its outcomes. *Paediatric and Perinatal Epidemiology, 20*, 148–162.

51. Looker, A. C., et al. (1997). Prevalency of iron deficiency in the United States. *Journal of the American Medical Association, 277*, 973–976.

52. Halterman, J. S., et al. (2001). Iron deficiency and cognitive achievement among school-aged children and adolescents in the United States. *Pediatrics, 107*, 1381–1386.

53. Lozoff, B., Jimenez, E., & Smith, J. B. (2006). Double burden of iron deficiency in infancy and low socioeconomic status: A longitudinal analysis of cognitive test scores to age 19 years. *Archives of Pediatrics and Adolescent Medicine, 160*, 1108–1113.

54. Tamura, T., et al. (2002). Cord serum ferritin concentrations and mental and psychomotor development of children at five years of age. *Journal of Pediatrics, 140*, 165–170.

55. McNeil, D. G. (2006). In raising the world's IQ, the secret's in the salt. New York: The New York Times

56. Tai, M. (1997). The devastating consequence of iodine deficiency. *Southeast Asian Journal of Tropical Medicine and Public Health, 28*(Suppl 2), 75–77.

57. Fenzi, G. F., et al. (1990). Neuropsychological assessment in schoolchildren from an area of moderate iodine deficiency. *Journal of Endocrinological Investigation, 13*, 427–431.

58. Aghini-Lombardi, F. A., et al. (1995). Mild iodine deficiency during fetal/neonatal life and neuropsychological impairment in Tuscany. *Journal of Endocrinological Investigation, 18*, 57–62.

59. Becker, D. V., et al. (2006). Iodine supplementation for pregnancy and lactation-United States and Canada: Recommendations of the American Thyroid Association. *Thyroid, 16*, 949–951.

60. Ray, J. G., et al. (2002). Association of neural tube defects and folic acid food fortification in Canada. *Lancet, 360*, 2047–2048.

61. Barbaux, S., Plomin, R., & Whitehead, A. S. (2000). Polymorphisms of genes controlling homocysteine/folate meatbolism and cognitive function. *Neuroreport, 11*, 1133–1136.

62. Gomez-Pinilla, F. (2008). Brain foods: The effects of nutrients on brain function. *Nature Reviews. Neuroscience, 9*, 568–578.

63. Malek, M. A., et al. (2006). Diarrhea- and rotavirus-associated hospitalization among children less than 5 years of age: United States, 1997 and 2000. *Pediatrics, 117*, 1887–1892.

64. Patrick, P. D., et al. (2005). Limitations in verbal fluency following heavy burdens of early childhood diarrhea in Brazilian shantytown children. *Child Neuropsychology, 11*, 233–244.

65. Lorntz, B., et al. (2006). Early childhood diarrhea predicts impaired school performance. *Pediatric Infectious Disease, 25*, 513–520.

66. Prado, M. S., et al. (2005). Asymptomatic giardiasis and growth in young children; a longitudinal study in Salvador, Brazil. *Parasitology, 131*, 51–56.

67. Muniz, P. T., et al. (2002). Intestinal parasitic infections in young children in Sao Paulo, Brazil: Prevalences, temporal trends and associations with physical growth. *Annals of Tropical Medicine and Parasitology, 96*, 503–512.

68. Simsek, Z., Zeyrek, F. Y., & Kurcer, M. A. (2004). Effect of Giardia infection on growth and psychomotor development of children aged 0–5 years. *Journal of Tropical Pediatrics, 50*, 90–93.

69. Celiksoz, A., et al. (2005). Effects of giardiasis on school success, weight and height indices of primary school children in Turkey. *Pediatrics International, 47*, 567–571.

70. Hamvas, A. (2000). Disparate outcomes for very low birth weight infants: genetics, environment, or both? *Journal of Pediatrics, 136*, 427–428.

71. Short, E. J., et al. (2003). Cognitive and academic consequences of bronchopulmonary dysplasia and very low birth weight: 8-year-old outcomes. *Pediatrics, 112*, e359–e366.

72. Hack, M., et al. (2002). Outcomes in young adulthood for very-low-birth-weight infants. *New England Journal of Medicine, 346*, 149–157.

73. Taylor, H. G., et al. (2000). Middle-school-age outcomes in children with very low birthweight. *Child Development, 71*, 1495–1511.

74. Hack, M., et al. (1994). School-age outcomes in children with birth weights under 750 g. *New England Journal of Medicine, 331*, 753–759.

75. Grunau, R. E., Whitfield, M. F., & Fay, T. B. (2004). Psychosocial and academic characteristics of extremely low birth weight ($\leq$ 800 g) adolescents who are free of major impairment compared with term-born control subjects. *Pediatrics, 114*, e725–e732.

76. National Center for Children in Poverty. (2006). Basic facts about low-income children: birth to age 18

77. Brooks-Gunn, J., Klebanov, P. K., & Duncan, G. J. (1996). Ethnic differences in children's intelligence test scores: role of economic deprivation, home enviroment, and maternal characteristics. *Child Development, 67*, 396–408.

78. Lupien, S. J., et al. (2001). Can poverty get under your skin? Basal cortisol levels and cognitive function in children from low and high socioeconomic status. *Development and Psychopathology, 13*, 653–676.

79. Smith, J. R., Brooks-Gunn, J., & Klebanov, P. K. (1997). Consequences of living in poverty for young children's cognitive and verbal ability and early school achievement. In G. J. Duncan & J. Brooks-Gunn (Eds.), *Consequences of Growing Up Poor* (pp. 132–189). New York: Russell Sage Foundation.

80. Feldman, M. A., & Walton-Allen, N. (1997). Effects of maternal mental retardation and poverty on intellectual, academic, and behavioral status of school-age children. *American Journal of Mental Retardation, 101*, 352–364.

81. Steen, R. G. (1996). *DNA & Destiny: Nature and Nurture in Human Behavior* (p. 295). New York: Plenum.

82. Turkheimer, E., et al. (2003). Socioeconomic status modifies heritability of IQ in young children. *Psychological Science, 14*, 623–628.

83. Kieffer, D. A., & Goh, D. S. (1981). The effect of individually contracted incentives on intelligence test performance of middle-and lower-SES children. *Journal of Clinical Psychology, 37*, 175–179.

84. Walker, S. P., et al. (2007). Child development: Risk factors for adverse outcomes in developing countries. *Lancet, 369*, 145–157.

85. Nelson, C. A., et al. (2007). Cognitive recovery in socially deprived young children: The Bucharest Early Intervention Project. *Science, 318*, 1937–1940.

86. Johnson, D. E., et al. (1992). The health of children adopted from Romania. *Journal of American Medical Association, 268*, 3446–3451.

87. Rutter, M. L., Kreppner, J. M., & O'Connor, T. G. (2001). Specificity and heterogeneity in children's responses to profound institutional privation. *British Journal of Psychiatry, 179*, 97–103.

88. Rutter, M. L., et al. (1998). Developmental catch-up and deficit following adoption after severe global early privation. *Journal of Child Psychology and Psychiatry, 39*, 465–476.

89. Walker, S. P., et al. (2005). Effects of early childhood psychosocial stimulation and nutritional supplementation on cognition in growth-stunted Jamaican children: Prospective cohort study. *Lancet, 366*, 1804–1807.

90. Yuan, W., et al. (2006). The impact of early childhood lead exposure on brain organization: A functional magnetic resonance imaging study of language function. *Pediatrics, 118*, 971–977.

# Chapter 8

1. Pinker, S., & Jackendoff, R. (2005). The faculty of language: what's special about it? *Cognition, 95*, 201–236.

2. Steen, R. G. (2007). *The Evolving Brain: The Known and the Unknown* (p. 437). New York: Prometheus Books.

3. Sattler, J. M. (1992). Wechsler Intelligence Scale for Children-Revised (WISC-R): Description, in *Assessment of Children* (p. 995). San Diego, CA: Jerome M. Sattler.

4. Marcus, G., & Rabagliati, H. (2006). What developmental disorders can tell us about the nature and origins of language. *Nature Neuroscience, 9*, 1226–1229.

5. Hurst, J. A., et al. (1990). An extended family with a dominantly inherited speech disorder. *Developmental Medicine and Child Neurology, 32*, 352–355.

6. Lai, C. S., et al. (2001). A forkhead-domain gene is mutated in a severe speech and language disorder. *Nature, 413*, 519–523.

7. Lai, C. S., et al. (2003). FOXP2 expression during brain development coincides with adult sites of pathology in a severe speech and language disorder. *Brain, 126*, 2455–2462.

8. Enard, W., et al. (2002). Molecular evolution of FOXP2, a gene involved in speech and language. *Nature, 418*, 869–872.

9. Fisher, S. E., & Marcus, G. F. (2006). The eloquent ape: genes, brain and the evolution of language. *Nature Reviews. Genetics, 7*, 9–20.

10. Bartlett, C. W., et al. (2002). A major susceptibility locus for specific language impairment is located on 13q21. *American Journal of Human Biology, 71*, 45–55.

11. Consortium, T. S. (2004). Highly significant linkage to the SLI1 locus in an expanded sample of individuals affected by specific language impairment. *American Journal of Human Biology , 74*, 1225–1238.

12. Ylisaukko-Oja, T., et al. (2005). Family-based association study of DYX1C1 variants in autism. *European Journal of Human Genetics, 13*, 127–130.
13. Steen, R. G. (1996). *DNA & Destiny: Nature and Nurture in Human Behavior* (p. 295). New York: Plenum Press.
14. Viding, E., et al. (2004). Genetic and environmental influence on language impairment in 4-year-old same-sex and opposite-sex twins. *Journal of Child Psychology and Psychiatry, and Allied Disciplines, 45*, 315–325.
15. Spinath, F. M., et al. (2004). The genetic and environmental origins of language disability and ability. *Child Development, 75*, 445–454.
16. Price, T. S., Dale, P. S., & Plomin, R. (2004). A longitudinal genetic analysis of low verbal and nonverbal cognitive abilities in early childhood. *Twin Research, 7*, 139–148.
17. Bishop, D. V., Adams, C. V., & Norbury, C. F. (2004). Using nonword repetition to distinguish genetic and environmental influences on early literacy development: a study of 6-year-old twins. *American Journal of Medical Genetics. Part B, Neuropsychiatric Genetics , 129*, 94–96.
18. DeThorne, L. S., et al. (2005). Low expressive vocabulary: higher heritability as a function of more severe cases. *Journal of Speech, Language, and Hearing Research, 48*, 792–804.
19. Bates, T. C., et al. (2007). Replication of reported linkages for dyslexia and spelling and suggestive evidence for novel regions on chromosomes 4 and 17. *European Journal of Human Genetics, 15*, 194–203.
20. Friederici, A. (2005). Neurophysiological markers of early language acquisition: from syllables to sentences. *Trends in Cognitive Sciences, 9*, 481–488.
21. Williford, J. A., Leech, S. L., & Day, N. L. (2006). Moderate prenatal alcohol exposure and cognitive status of children at age 10. *Alcoholism, Clinical and Experimental Research, 30*, 1051–1059.
22. Howell, K. K., et al. (2005). Prenatal alcohol exposure and ability, academic achievement, and school functioning in adolescence: a longitudinal follow-up. *Journal of Pediatric Psychology, 31*, 116–126.
23. Jacobson, S. W., et al. (2004). Maternal age, alcohol abuse history, and quality of parenting as moderators of the effects of prenatal alcohol exposure on 7.5-year intellectual function. *Alcoholism, Clinical and Experimental Research, 28*, 1732–1745.
24. Viljoen, D. L., et al. (2005). Fetal alcohol syndrome epidemiology in a South African community: a second study of a very high prevalence area. *Journal of Studies on Alcohol, 66*, 593–604.
25. May, P. A., et al. (2006). Epidemiology of FASD in a province in Italy: prevalence and characteristics of children in a random sample of schools. *Alcoholism, Clinical and Experimental Research, 30*, 1562–1575.
26. Jacobson, S. W., et al. (2002). Validity of maternal report of prenatal alcohol, cocaine, and smoking in relation to neurobehavioral outcome. *Pediatrics, 109*, 815–825.
27. Wozniak, J. R., et al. (2006). Diffusion tensor imaging in children with fetal alcohol spectrum disorders. *Alcoholism, Clinical and Experimental Research, 30*, 1799–1806.
28. Astley, S. J., et al. (2000). Fetal alcohol syndrome (FAS) primary prevention through FAS diagnosis: II. A comprehensive profile of 80 birth mothers of children with FAS. *Alcohol and Alcoholism, 35*, 509–519.
29. Braudel, F. (1979). *The Structures of Everyday Life: The Limits of the Possible. Civilization & Capitalism, 15th-18th Century* (Vol. 1, p. 622). New York: Harper & Row.

# Chapter 9

1. United States Census Bureau. (2006). American community survey data profile highlights. In *American factfinder*. Washington, DC: US Census Bureau.
2. Stiers, P., et al. (2002). Visual-perceptual impairment in a random sample of children with cerebral palsy. *Developmental Medicine and Child Neurology, 44*, 370–382.

3. Editorial. (2007). Counting the poor. In *New York Times*. New York.

4. Sampson, R. J., Sharkey, P., & Raudenbush, S. W. (2008). Durable effects of concentrated disadvantage on verbal ability among African-American children. *Proceedings of the National Academy of Sciences of the United States of America, 105*(3), 845–852.

5. Wardlaw, T., et al. (2004). *Low birthweight: Country, regional and global estimates* (p. 31). Geneva: UNICEF and WHO.

6. Short, E. J., et al. (2003). Cognitive and academic consequences of bronchopulmonary dysplasia and very low birth weight: 8-year-old outcomes. *Pediatrics, 112*, e359–e366.

7. Kessler, R. C., et al. (2005). Lifetime prevalence and age-at-onset distributions of DSM-IV disorders in the National Comorbidity Survey Replication. *Archives of General Psychiatry, 62*, 593–602.

8. Andreou, C., et al. (2003). Verbal intelligence and sleep disorders in children with ADHD. *Perceptual and Motor Skills, 96*, 1283–1288.

9. Oddy, W. H., et al. (2003). Breast feeding and cognitive development in childhood: A prospective birth cohort study. *Paediatric and Perinatal Epidemiology, 17*(1), 81–90.

10. Williford, J. A., Leech, S. L., & Day, N. L. (2006). Moderate prenatal alcohol exposure and cognitive status of children at age 10. *Alcoholism, Clinical and Experimental Research, 30*, 1051–1059.

11. Centers for Disease Control. (2006). Use of cigarettes and other tobacco products among students aged 13–15 years worldwide, 1999–2005. *Morbidity and Mortality Weekly Report, 55*(20), 553–556.

12. Fried, P. A., Watkinson, B., & Gray, R. (2006). Neurocognitive consequences of cigarette smoking in young adults – a comparison with pre-drug performance. *Neurotoxicology and Teratology, 28*, 517–525.

13. Lanphear, B. P., et al. (2005). Low-level environmental lead exposure and children's intellectual function: An international pooled analysis. *Environmental Health Perspectives, 113*, 894–899.

14. Canfield, R. L., et al. (2003). Intellectual impairment in children with blood lead concentrations below 10 ug per deciliter. *The New England Journal of Medicine, 348*, 1517–1526.

15. Tattersfield, A. E., et al. (2002). Asthma. *Lancet, 360*, 1313–1322.

16. Bacharier, L. B., et al. (2003). Hospitalization for asthma: Atopic, pulmonary function, and psychological correlates among participants in the Childhood Asthma Management Program. *Pediatrics, 112*, e85–e92.

17. Stewart, M. G. (2008). Identification and management of undiagnosed and undertreated allergic rhinitis in adults and children. *Clinical and Experimental Allergy, 38*, 751–760.

18. Bender, B. G. (2005). Cognitive effects of allergic rhinitis and its treatment. *Immunology and Allergy Clinics of North America, 25*, 301–312.

19. Wilken, J. A., Berkowitz, R., & Kane, R. (2002). Decrements in vigilance and cognitive functioning associated with ragweed-induced allergic rhinitis. *Annals of Allergy, Asthma and Immunology, 89*, 372–380.

20. Vuurman, E. F., et al. (1993). Seasonal allergic rhinitis and antihistamine effects on children's learning. *Annals of Allergy, Asthma and Immunology, 71*, 121–126.

21. Kremer, B., den Hartog, H. M., & Jolles, J. (2002). Relationship between allergic rhinitis, disturbed cognitive functions and psychological well-being. *Clinical and Experimental Allergy, 32*, 1310–1315.

22. McCarthy, J., et al. (2004). Sustained attention and visual processing speed in children and adolescents with bipolar disorder and other psychiatric disorders. *Psychological Reports, 95*, 39–47.

23. Saltzman, K. M., Weems, C. F., & Carrion, V. G. (2006). IQ and posttraumatic stress symptoms in children exposed to interpersonal violence. *Child Psychiatry and Human Development, 36*, 261–272.

24. Bremner, J. D., et al. (2004). Deficits in verbal declarative memory function in women with childhood sexual abuse-related posttraumatic stress disorder. *The Journal of Nervous and Mental Disease, 192*, 643–649.

25. Goodwin, R. D., & Stein, M. B. (2004). Association between childhood trauma and physical disorders among adults in the United States. *Psychological Medicine, 34*, 509–520.

26. Nelson, C. A., et al. (2007). Cognitive recovery in socially deprived young children: The Bucharest Early Intervention Project. *Science, 318*, 1937–1940.

27. Gallagher, P., et al. (2007). Neurocognitive function following remission in major depression disorder: Potential objective marker of response? *The Australian and New Zealand Journal of Psychiatry, 41*, 54–61.

28. Xu, Z., Cheuk, D. K. L., & Lee, S. L. (2006). Clinical evaluation in predicting childhood obstructive sleep apnea. *Chest, 130*, 1765–1771.

29. Halbower, A. C., et al. (2006). Childhood obstructive sleep apnea associates with neuropsychological deficits and neuronal brain injury. *PLoS Medicine, 3*, e301.

30. Centers for Disease Control (2002). Iron deficiency – United States, 1999–2000. In *MMWR, Morbidity and Mortality Weekly Report*. Atlanta, GA: CDC.

31. Lozoff, B., Jimenez, E., & Smith, J. B. (2006). Double burden of iron deficiency in infancy and low socioeconomic status: A longitudinal analysis of cognitive test scores to age 19 years. *Archives of Pediatrics and Adolescent Medicine, 160*, 1108–1113.

32. Tamura, T., et al. (2002). Cord serum ferritin concentrations and mental and psychomotor development of children at five years of age. *The Journal of Pediatrics, 140*, 165–170.

33. Carpenter, D. O. (2006). Polychlorinated biphenyls (PCBs): Routes of exposure and effects on human health. *Reviews on Environmental Health, 21*, 1–23.

34. Everett, C. J., et al. (2007). Association of a polychlorinated dibenzo-p-dioxin, a polychlorinated biphenyl, and DDT with diabetes in the 1999–2002 National Health and Nutrition Examination Survey. *Environmental Research, 103*, 413–418.

35. Jacobson, J. L., & Jacobson, S. W. (1996). Intellectual impairment in children exposed to polychlorinated biphenyls *in utero*. *The New England Journal of Medicine, 335*, 783–789.

36. Newman, J., et al. (2006). PCBs and cognitive functioning of Mohawk adolescents. *Neurotoxicology and Teratology, 28*, 439–445.

37. Patandin, S., et al. (1999). Effects of environmental exposure to polychlorinated biphenyls and dioxins on cognitive abilities in Dutch children at 42 months of age. *The Journal of Pediatrics, 134*, 33–41.

38. Gray, K. A., et al. (2005). *In utero* exposure to background levels of polychlorinated biphenyls and cognitive functioning among school-age children. *American Journal of Epidemiology, 162*, 17–26.

39. Wild, S., et al. (2004). Global prevalence of diabetes: Estimates for the year 2000 and projections for 2030. *Diabetes Care, 27*(5), 1047–1053.

40. McCarthy, A. M., et al. (2002). Effects of diabetes on learning in children. *Pediatrics, 109*, E9–E19.

41. Wu, W., et al. (2008). The brain in the age of old: The hippocampal formation is targeted differentially by diseases of late life. *Annals of Neurology, 64*, 698–706.

42. Kaufman, F. R., et al. (1999). Neurocognitive functioning in children diagnosed with diabetes before age 10 years. *Journal of Diabetes and its Complications, 13*, 31–38.

43. Northam, E. A., et al. (2001). Neuropsychological profiles of children with Type I diabetes 6 years after disease onset. *Diabetes Care, 24*, 1541–1546.

44. Koenen, K. C., et al. (2003). Domestic violence is associated with environmental suppression of IQ in young children. *Development and Psychopathology, 15*, 297–311.

45. Choi, H., et al. (2008). Prenatal exposure to airborne polycyclic aromatic hydrocarbons and risk of intrauterine growth restriction. *Environmental Health Perspectives, 116*, 658–665.

46. Perera, F., et al. (2006). Effect of prenatal exposure to airborne polycyclic aromatic hydrocarbons on neurodevelopment in the first 3 years of life among inner-city children. *Environmental Health Perspectives, 114*, 1287–1292.

47. Vasa, R. A., et al. (2006). Memory deficits in children with and at risk for anxiety disorders. *Depression and Anxiety, 24*, 85–94.

48. Schober, S. E., et al. (2003). Blood mercury levels in US children and women of childbearing age, 1999–2000. *Journal of the American Medical Association, 289*, 1667–1674.

49. Trasande, L., Landrigan, P. J., & Schechter, C. (2005). Public health and economic consequences of methyl mercury toxicity to the developing brain. *Environmental Health Perspectives, 113*, 590–596.

50. Cohen, J. T., Bellinger, D. C., & Shaywitz, B. A. (2005). A quantitative analysis of prenatal methyl mercury exposure and cognitive development. *American Journal of Preventive Medicine, 29*, 353–365.

51. Koman, L. A., Smith, B. P., & Shilt, J. S. (2004). Cerebral palsy. *Lancet, 363*, 1619–1631.

52. MacDonald, B. K., et al. (2000). The incidence and lifetime prevalence of neurological disorders in a prospective community-based study in the UK. *Brain, 123*, 665–676.

53. Leary, P. M., et al. (1999). Childhood secondary (symptomatic) epilepsy, seizure control, and intellectual handicap in a nontropical region of South Africa. *Epilepsia, 40*, 1110–1113.

54. O'Leary, S. D., Burns, T. G., & Borden, K. A. (2006). Performance of children with epilepsy and normal age-matched controls on the WISC-III. *Child Neuropsychology, 12*, 173–180.

55. Mathers, C., Smith, A. & Concha, M. (2000). Global burden of hearing loss in the year 2000. In WHO (Ed.), *Global burden of disease*. Geneva: WHO.

56. Lindsay, R. L., et al. (1999). Early ear problems and developmental problems at school age. *Clinical Pediatrics, 38*, 123–132.

57. Wake, M., et al. (2005). Hearing impairment: A population study of age at diagnosis, severity, and language outcomes at 7–8 years. *Archives of Disease in Childhood, 90*, 238–244.

58. Admiraal, R. J., & Huygen, P. L. (1999). Causes of hearing impairment in deaf pupils with a mental handicap. *International Journal of Pediatric Otorhinolaryngology, 51*, 101–108.

59. Vernon, M. (2005). Fifty years of research on the intelligence of deaf and hard-of-hearing children: A review of literature and discussion of implications. *Journal of Deaf Studies and Deaf Education, 10*, 225–231.

60. Christianson, A., Howson, C. P. & Modell, B. (2006) *Global report on birth defects: The hidden toll of dying and disabled children*. (pp. 1–18) White Plains: March of Dimes Birth Defects Foundation.

61. Forbess, J. M., et al. (2002). Neurodevelopmental outcome after congenital heart surgery: Results from an institutional registry. *Circulation, 106*(Suppl. I), I95–I102.

62. Schneider, H. J., et al. (2007). Hypopituitarism. *Lancet, 369*, 1461–1470.

63. Brown, K., et al. (2004). Abnormal cognitive function in treated congenital hypopituitarism. *Archives of Disease in Childhood, 89*, 827–830.

64. Christiani, D. (2003) Arsenic exposure and human health. In *Superfund basic research program*. Berkley,CA: National Institute of Environmental Health Sciences.

65. Calderon, J., et al. (2001). Exposure to arsenic and lead and neuropsychological development in Mexican children. *Environmental Research Section A, 85*, 69–76.

66. Wasserman, G. A., et al. (2004). Water arsenic exposure and children's intellectual function in Arailhazar Bangladesh. *Environmental Health Perspectives, 112*, 1329–1333.

67. Boswell, J. E., McErlean, M., & Verdile, V. P. (2002). Prevalence of traumatic brain injury in an ED population. *The American Journal of Emergency Medicine, 20*, 177–180.

68. Hawley, C. A., et al. (2004). Return to school after brain injury. *Archives of Disease in Childhood, 89*, 136–142.

69. Ewing-Cobbs, L., et al. (2006). Late intellectual and academic outcomes following traumatic brain injury sustained during early childhood. *Journal of Neurosurgery, 105*, 287–296.

70. Hogan, A. M., et al. (2006). Physiological correlates of intellectual function in children with sickle cell disease: Hypoxaemia, hyperaemia and brain infarction. *Developmental Science, 9*, 379–387.

71. Steen, R. G., et al. (2005). Cognitive deficits in children with sickle cell disease. *Journal of Child Neurology, 20*, 102–107.

72. National Institute of Allergy and Infectious Diseases. (2004). HIV infection in infants and children. In *NIAID fact sheet*. Washington, DC: National Institutes of Health.

73. Nozyce, M. L., et al. (2006). A behavioral and cognitive profile of clinically stable HIV-infected children. *Pediatrics, 117*, 763–770.

74. Taha, T. E., et al. (2004). Nevirapine and zidovudine at birth to reduce perinatal transmission of HIV in an African setting: A randomized controlled trial. *The Journal of the American Medical Association, 292,* 202–209.

75. Robbins, J. M., et al. (2006). Reduction in newborns with discharge coding of *in utero* alcohol effects in the United States, 1993 to 2002. *Archives of Pediatrics and Adolescent Medicine, 160*(12), 1224–1231.

76. Steen, R. G. & Campbell, F. A. (2008) The cognitive impact of systemic illness in childhood and adolescence. In L. Weiss & A. Prifitera (Eds.), *WISC-IV: Clinical Use and Interpretation* (pp. 365–407). New York: Academic Press.

77. Campbell, F. A., et al. (2001). The development of cognitive and academic abilities: Growth curves from an early childhood educational experiment. *Developmental Psychology, 37,* 231–242.

78. Brosco, J. P., Mattingly, M., & Sanders, L. M. (2006). Impact of specific medical interventions on reducing the prevalence of mental retardation. *Archives of Pediatrics and Adolescent Medicine, 160,* 318–320.

79. Flynn, J. (1984). The mean IQ of Americans: Massive gains 1932 to 1978. *Psychological Bulletin, 95,* 29–51.

80. Flynn, J. R. (1987). Massive IQ gains in 14 nations: What IQ tests really measure. *Psychological Bulletin, 101,* 171–191.

81. Flynn, J. R. (1998). Israeli military IQ tests: Gender differences small; IQ gains large. *Journal of Biosocial Science, 30,* 541–553.

# Chapter 10

1. United Nations. (2007). *World Population Prospects: The 2006 Revision.* Population Division, Population Estimates and Projections Section, Population Division, Editor. Dept. Economic and Social Affairs.

2. Idro, R., et al. (2006). Risk factors for persisting neurological and cognitive impairments following cerebral malaria. *Archives of Disease in Childhood, 91,* 142–148.

3. Fernando, S. D., et al. (2003). The impact of repeated malaria attacks on the school performance of children. *The American Journal of Tropical Medicine and Hygiene, 69,* 582–588.

4. Fernando, D., et al. (2006). A randomized, double-blind, placebo-controlled, clinical trial of the impact of malaria prevention on the educational attainment of school children. *The American Journal of Tropical Medicine and Hygiene, 74,* 386–393.

5. Gamble, C., et al. (2007). Insecticide-treated nets for the prevention of malaria in pregnancy: a systematic review of randomised clontrolled trials. *PLoS Medicine, 4,* E107.

6. Walker, S. P., et al. (2007). Child development: risk factors for adverse outcomes in developing countries. *Lancet, 369,* 145–157.

7. Hibbeln, J. R., et al. (2007). Maternal seafood consumption in pregnancy and neurodevelopmental outcomes in childhood (ALSPAC study): an observational cohort study. *Lancet, 369,* 578–585.

8. Oerbeck, B., et al. (2003). Congenital hypothyroidism: influence of disease severity and L-thyroxine treatment on intellectual, motor, and school-associated outcomes in young adults. *Pediatrics, 112,* 923–930.

9. Zimmermann, M. B., & Hurrell, R. F. (2007). Nutritional iron deficiency. *Lancet, 370,* 511–520.

10. Lozoff, B., Jimenez, E., & Smith, J. B. (2006). Double burden of iron deficiency in infancy and low socioeconomic status: a longitudinal analysis of cognitive test scores to age 19 years. *Archives of Pediatrics & Adolescent Medicine, 160,* 1108–1113.

11. Tamura, T., et al. (2002). Cord serum ferritin concentrations and mental and psychomotor development of children at five years of age. *The Journal of Pediatrics, 140,* 165–170.

12. Fawzi, W. W., et al. (2007). Vitamins and perinatal outcomes among HIV-negative women in Tanzania. *The New England Journal of Medicine, 356*, 1423–1431.
13. Hotez, P. J., et al. (2005). Hookworm: "The great infection of mankind". *PLoS Medicine, 2*, e67.
14. Brooker, S., et al. (2007). Age-related changes in hookworm infection, anaemia and iron deficiency in an area of high *Necator americanus* hookworm transmission in south-eastern Brazil. *Transactions of the Royal Society of Tropical Medicine and Hygiene, 101*, 146–154.
15. Sakti, H., et al. (1999). Evidence for an association between hookworm infection and cognitive function in Indonesian school children. *Tropical Medicine & International Health, 4*, 322–334.
16. Reddy, M., et al. (2007). Oral drug therapy for multiple neglected tropical diseases: a systematic review. *The Journal of the American Medical Association, 298*, 1911–1924.
17. Brooker, S., et al. (2000). Epidemiology of single and multiple species of helminth infections among school children in Busia District, Kenya. *East African Medical Journal, 77*, 157–161.
18. Ezeamama, A. E., et al. (2005). Helminth infection and cognitive impairment among Filipino children. *The American Journal of Tropical Medicine and Hygiene, 72*, 540–548.
19. Hadidjaja, P., et al. (1998). The effect of intervention methods on nutritional status and cognitive function of primary school children infected with *Ascaris lumbricoides*. *The American Journal of Tropical Medicine and Hygiene, 59*(5), 791–795.
20. Ross, A. G. P., et al. (2002). Schistosomiasis. *The New England Journal of Medicine, 346*, 1212–1220.
21. Nokes, C., et al. (1999). Evidence for an improvement in cognitive function following treatmenr of *Schistosoma japonicaum* infection in Chinese primary schoolchildren. *The American Journal of Tropical Medicine and Hygiene, 60*, 556–565.
22. Jukes, M. C., et al. (2002). Heavy schistosomiasis associated with poor short-term memory and slower reaction times in Tanzanian schoolchildren. *Tropical Medicine & International Health, 7*, 104–117.
23. Steer, P. (2005). The epidemiology of preterm labour. *British Journal of Obstetrics and Gynaecology, 112*(Suppl 1), 1–3.
24. Peterson, B. S., et al. (2000). Regional brain volume abnormalities and long-term cognitive outcome in preterm infants. *The Journal of the American Medical Association, 284*, 1939–1947.
25. Bhutta, A. T., et al. (2002). Cognitive and behavioral outcomes of school-aged children who were born preterm: a meta-analysis. *The Journal of the American Medical Association, 288*, 728–737.
26. Rappley, M. D. (2005). Attention deficit-hyperactivity disorder. *The New England Journal of Medicine, 352*, 165–173.
27. Quinlan, D. M., & Brown, T. E. (2003). Assessment of short-term verbal memory impairments in adolescents and adults with ADHD. *Journal of Attention Disorders, 6*, 143–152.
28. Andreou, C., et al. (2003). Verbal intelligence and sleep disorders in children with ADHD. *Perceptual and Motor Skills, 96*, 1283–1288.
29. Montoya, J. G., & Liesenfeld, O. (2004). Toxoplasmosis. *Lancet, 363*, 1965–1976.
30. Thiebaut, R., & Group, S. S. (2007). Effectiveness of prenatal treatment for congenital toxoplasmosis: a meta-analysis of individual patients' data. *Lancet, 369*, 115–122.
31. Roizen, N., et al. (2006). Impact of visual impairment on measures of cognitive function for children with congenital toxoplasmosis: implications for compensatory intervention strategies. *Pediatrics, 118*, e379.
32. Kastrup, M. C., & Baez Ramos, A. (2007). Global mental health. *Danish Medical Bulletin, 54*, 42–43.
33. Saltzman, K. M., Weems, C. F., & Carrion, V. G. (2006). IQ and posttraumatic stress symptoms in children exposed to interpersonal violence. *Child Psychiatry and Human Development, 36*, 261–272.
34. Bremner, J. D., et al. (2004). Deficits in verbal declarative memory function in women with childhood sexual abuse-related posttraumatic stress disorder. *The Journal of Nervous and Mental Disease, 192*, 643–649.

35. Goodwin, R. D., & Stein, M. B. (2004). Association between childhood trauma and physical disorders among adults in the United States. *Psychological Medicine, 34*, 509–520.
36. Walker, S. P., et al. (2005). Effects of early childhood psychosocial stimulation and nutritional supplementation on cognition in growth-stunted Jamaican children: prospective cohort study. *Lancet, 366*, 1804–1807.
37. Nelson, C. A., et al. (2007). Cognitive recovery in socially deprived young children: the Bucharest Early Intervention Project. *Science, 318*, 1937–1940.
38. Simeon, D. T., et al. (1995). Treatment of *Trichuris trichiura* infections improves growth, spelling scores and school attendance in some children. *The Journal of Nutrition, 125*, 1875–1883.
39. de Onis, M., et al. (2004). Methodology for estimating regional and global trends of child malnutrition. *International Journal of Epidemiology, 33*, 1–11.
40. Lucas, A., Morley, R., & Cole, T. J. (1998). Randomised trial of early diet in preterm babies and later intelligence quotient. *British Medical Journal, 317*, 1481–1487.
41. Liu, J., et al. (2003). Malnutrition at age 3 years and lower cognitive ability at age 11 years: independence from psychosocial adversity. *Archives of Pediatrics & Adolescent Medicine, 157*, 593–600.
42. Edlich, R. F., et al. (2005). Rubella and congenital rubella (German measles). *Journal of Long-Term Effects of Medical Implants, 15*, 319–328.
43. Givens, K. T., et al. (1993). Congenital rubella syndrome: ophthalmic manifestations and associated systemic disorders. *The British Journal of Ophthalmology, 77*, 358–363.
44. Watts, J. (2005). China: the air pollution capital of the world. *Lancet, 366*, 1761–1762.
45. Hopkins, R. O., & Woon, F. L. (2006). Neuroimaging, cognitive, and neurobehavioral outcomes following carbon monoxide poisoning. *Behavioral and Cognitive Neuroscience Reviews, 5*, 141–155.
46. Amitai, Y., et al. (1998). Neuropsychological impairment from acute low-level exposure to carbon monoxide. *Archives of Neurology, 55*, 845–848.
47. Perera, F., et al. (2006). Effect of prenatal exposure to airborne polycyclic aromatic hydrocarbons on neurodevelopment in the first 3 years of life among inner-city children. *Environmental Health Perspectives, 114*, 1287–1292.
48. Tang, D., et al. (2006). PAH-DNA adducts in cord blood and fetal and child development in a Chinese cohort. *Environmental Health Perspectives, 114*, 1297–1300.
49. Wild, S., et al. (2004). Global prevalence of diabetes: estimates for the year 2000 and projections for 2030. *Diabetes Care, 27*(5), 1047–1053.
50. McCarthy, A. M., et al. (2002). Effects of diabetes on learning in children. *Pediatrics, 109*, E9–E19.
51. Kaufman, F. R., et al. (1999). Neurocognitive functioning in children diagnosed with diabetes before age 10 years. *J Diabetes Complications, 13*, 31–38.
52. Northam, E. A., et al. (2001). Neuropsychological profiles of children with Type I diabetes 6 years after disease onset. *Diabetes Care, 24*, 1541–1546.
53. Neggers, Y. H., et al. (2003). Maternal prepregnancy body mass index and psychomotor development in children. *Acta Obstetricia et Gynecologica Scandinavica, 82*, 235–240.
54. Diabetes Control and Complications Trial. (2007). Long-term effects of diabetes and its treatment on cognitive function. *The New England Journal of Medicine, 356*, 1842–1852.
55. Williford, J. A., Leech, S. L., & Day, N. L. (2006). Moderate prenatal alcohol exposure and cognitive status of children at age 10. *Alcoholism, Clinical and Experimental Research, 30*, 1051–1059.
56. Centers for Disease Control. (2006). Use of cigarettes and other tobacco products among students aged 13–15 years worldwide, 1999–2005. *MMWR Weekly, 55*(20), 553–556.
57. Fried, P. A., Watkinson, B., & Gray, R. (2006). Neurocognitive consequences of cigarette smoking in young adults – a comparison with pre-drug performance. *Neurotoxicology and Teratology, 28*, 517–525.
58. Patel, V., et al. (2001). Depression in developing countries: lessons from Zimbabwe. *British Medical Journal, 322*, 482–484.

59. Gallagher, P., et al. (2007). Neurocognitive function following remission in major depression disorder: potential objective marker of response? *The Australian and New Zealand Journal of Psychiatry, 41*, 54–61.

60. Marconi, V. C., Garcia, H. H., & Katz, J. T. (2006). Neurocysticercosis. *Current Infectious Disease Reports, 8*, 293–300.

61. Terra-Bustamante, V. C., et al. (2005). Cognitive performance of patients with mesial temporal lobe epilepsy and incidental calcified neurocysticercosis. *Journal of Neurology, Neurosurgery, and Psychiatry, 76*, 1080–1083.

62. Pruss-Ustun, A., et al. (2004). Lead exposure. In M. Ezzati (Ed.), *Comparative quantification of health risks: global and regional burden of disease attributable to selected major risk factors.* Geneva: World Health Org.

63. Canfield, R. L., et al. (2003). Intellectual impairment in children with blood lead concentrations below 10 µg per deciliter. *The New England Journal of Medicine, 348*, 1517–1526.

64. Wasserman, G. A., et al. (2003). The relationship between blood lead, bone lead and child intelligence. *Child Neuropsychology, 9*, 22–34.

65. Horwood, L. J., Darlow, B. A., & Mogridge, N. (2001). Breast milk feeding and cognitive ability at 7–8 years. *Archives of Disease in Childhood. Fetal and Neonatal Edition, 84*, F23–F27.

66. Oddy, W. H., et al. (2003). Breast feeding and cognitive development in childhood: a prospective birth cohort study. *Paediatric and Perinatal Epidemiology, 17*(1), 81–90.

67. Koenen, K. C., et al. (2003). Domestic violence is associated with environmental suppression of IQ in young children. *Development and Psychopathology, 15*, 297–311.

68. Koman, L. A., Smith, B. P., & Shilt, J. S. (2004). Cerebral palsy. *Lancet, 363*, 1619–1631.

69. Stiers, P., et al. (2002). Visual-perceptual impairment in a random sample of children with cerebral palsy. *Developmental Medicine and Child Neurology, 44*, 370–382.

70. Nozyce, M. L., et al. (2006). A behavioral and cognitive profile of clinically stable HIV-infected children. *Pediatrics, 117*, 763–770.

71. Taha, T. E., et al. (2004). Nevirapine and zidovudine at birth to reduce perinatal transmission of HIV in an African setting: a randomized controlled trial. *The Journal of the American Medical Association, 292*, 202–209.

72. Xu, Z., Cheuk, D. K. L., & Lee, S. L. (2006). Clinical evaluation in predicting childhood obstructive sleep apnea. *Chest, 130*, 1765–1771.

73. Halbower, A. C., et al. (2006). Childhood obstructive sleep apnea associates with neuropsychological deficits and neuronal brain injury. *PLoS Medicine, 3*, e301.

74. Patrick, P. D., et al. (2005). Limitations in verbal fluency following heavy burdens of early childhood diarrhea in Brazilian shantytown children. *Child Neuropsychology, 11*, 233–244.

75. Niehaus, M. D., et al. (2002). Early childhood diarrhea is associated with diminished cognitive function 4 to 7 years later in children in a northeast Brazilian shantytown. *The American Journal of Tropical Medicine and Hygiene, 66*, 590–593.

76. Berkman, D. S., et al. (2002). Effects of stunting, diarrhoeal disease, and parasitic infection during infancy on cognition in late childhood: a follow-up study. *Lancet, 359*, 564–571.

77. Njenga, S. M., et al. (2007). Chronic clinical manifestations related to *Wucheria bancrofti* infection in a highly endemic area in Kenya. *Transactions of the Royal Society of Tropical Medicine and Hygiene, 101*, 439–444.

78. Haddix, A. C., & Kestler, A. (2000). Lymphatic filariasis: economic aspects of the disease and programmes for its elimination. *Transactions of the Royal Society of Tropical Medicine and Hygiene, 94*, 592–593.

79. Dreyer, G., et al. (2000). Pathogenesis of lymphatic disease in bancroftian filariasis: a clinical perspective. *Parasitology Today, 16*(12), 544–549.

80. Tattersfield, A. E., et al. (2002). Asthma. *Lancet, 360*, 1313–1322.

81. Bacharier, L. B., et al. (2003). Hospitalization for asthma: atopic, pulmonary function, and psychological correlates among participants in the Childhood Asthma Management Program. *Pediatrics, 112*, e85–e92.

82. Annett, R. D., Bender, B. G., & Gordon, M. (2007). Relating children's attentional capabilities to intelligence, memory, and academic achievement: a test of construct specificity in children with asthma. *Child Neuropsychology, 13*, 64–85.

83. Stewart, M. G. (2008). Identification and management of undiagnosed and undertreated allergic rhinitis in adults and children. *Clinical and Experimental Allergy, 38*, 751–760.

84. Bender, B. G. (2005). Cognitive effects of allergic rhinitis and its treatment. *Immunology and Allergy Clinics of North America, 25*, 301–312.

85. Wilken, J. A., Berkowitz, R., & Kane, R. (2002). Decrements in vigilance and cognitive functioning associated with ragweed-induced allergic rhinitis. *Annals of Allergy, Asthma & Immunology, 89*, 372–380.

86. Kapaj, S., et al. (2006). Human health effects from chronic arsenic poisoning – a review. *Journal of Environmental Science and Health. Part A, Toxic/Hazardous Substances & Environmental Engineering, 41*, 2399–2428.

87. Yu, G., Sun, D., & Zheng, Y. (2007). Health effects of exposure to natural arsenic in ground water and coal in China: an overview of occurrence. *Environmental Health Perspectives, 115*(4), 636–642.

88. Calderon, J., et al. (2001). Exposure to arsenic and lead and neuropsychological development in Mexican children. *Environmental Research Sect A, 85*, 69–76.

89. Wasserman, G. A., et al. (2004). Water arsenic exposure and children's intellectual function in Arailhazar, Bangladesh. *Environmental Health Perspectives, 112*, 1329–1333.

90. Kennedy, C. R., & W.U.H.H.S.T. Group. (1998). Controlled trial of universal neonatal screening for early identification of permanent childhood hearing impairment. *Lancet, 352*, 1957–1964.

91. Lindsay, R. L., et al. (1999). Early ear problems and developmental problems at school age. *Clinical Pediatrics (Phila), 38*, 123–132.

92. Wake, M., et al. (2005). Hearing impairment: a population study of age at diagnosis, severity, and language outcomes at 7–8 years. *Archives of Disease in Childhood, 90*, 238–244.

93. Admiraal, R. J., & Huygen, P. L. (1999). Causes of hearing impairment in deaf pupils with a mental handicap. *International Journal of Pediatric Otorhinolaryngology, 51*, 101–108.

94. Vernon, M. (2005). Fifty years of research on the intelligence of deaf and hard-of-hearing children: a review of literature and discussion of implications. *Journal of Deaf Studies and Deaf Education, 10*, 225–231.

95. Kennedy, C. R., et al. (2006). Language ability after early detection of permanent childhood hearing impairment. *The New England Journal of Medicine, 354*, 2131–2141.

96. MacDonald, B. K., et al. (2000). The incidence and lifetime prevalence of neurological disorders in a prospective community-based study in the UK. *Brain, 123*, 665–676.

97. Leary, P. M., et al. (1999). Childhood secondary (symptomatic) epilepsy, seizure control, and intellectual handicap in a nontropical region of South Africa. *Epilepsia, 40*, 1110–1113.

98. O'Leary, S. D., Burns, T. G., & Borden, K. A. (2006). Performance of children with epilepsy and normal age-matched controls on the WISC-III. *Child Neuropsychology, 12*, 173–180.

99. Wardlaw, T., et al. (2004). *Low birthweight: country, regional and global estimates* (p. 31). Geneva: UNICEF and WHO.

100. Short, E. J., et al. (2003). Cognitive and academic consequences of bronchopulmonary dysplasia and very low birth weight: 8-year-old outcomes. *Pediatrics, 112*, e359–e366.

101. Carpenter, D. O. (2006). Polychlorinated biphenyls (PCBs): routes of exposure and effects on human health. *Reviews on Environmental Health, 21*, 1–23.

102. Everett, C. J., et al. (2007). Association of a polychlorinated dibenzo-p-dioxin, a polychlorinated biphenyl, and DDT with diabetes in the 1999–2002 National Health and Nutrition Examination Survey. *Environmental Research, 103*, 413–418.

103. Jacobson, J. L., & Jacobson, S. W. (1996). Intellectual impairment in children exposed to polychlorinated biphenyls *in utero*. *The New England Journal of Medicine, 335*, 783–789.

104. Newman, J., et al. (2006). PCBs and cognitive functioning of Mohawk adolescents. *Neurotoxicology and Teratology, 28*, 439–445.

105. Patandin, S., et al. (1999). Effects of environmental exposure to polychlorinated biphenyls and dioxins on cognitive abilities in Dutch children at 42 months of age. *The Journal of Pediatrics, 134*, 33–41.

106. Gray, K. A., et al. (2005). *In utero* exposure to background levels of polychlorinated biphenyls and cognitive functioning among school-age children. *American Journal of Epidemiology, 162,* 17–26.

107. Druet-Cabanac, M., et al. (1999). Onchocerciasis and epilepsy: a matched case-control study in the Central African Republic. *American Journal of Epidemiology, 149,* 565–570.

108. Druet-Cabanac, M., et al. (2004). Review of epidemiological studies searching for a relationship between onchocerciasis and epilepsy. *Neuroepidemiology, 23*(3), 144–149.

109. Sloan, F. A., et al. (2005). Effects of change in self-reported vision on cognitive, affective, and functional status and living arrangements among the elderly. *American Journal of Ophthalmology, 140,* 618–627.

110. Christianson, A., Howson, C. P., & Modell, B. (2006). *Global report on birth defects: the hidden toll of dying and disabled children* (pp. 1–18). White Plains: March of Dimes Birth Defects Foundation.

111. Hogan, A. M., et al. (2006). Physiological correlates of intellectual function in children with sickle cell disease: hypoxaemia, hyperaemia and brain infarction. *Developmental Science, 9,* 379–387.

112. Steen, R. G., et al. (2005). Cognitive deficits in children with sickle cell disease. *Journal of Child Neurology, 20,* 102–107.

113. Schober, S. E., et al. (2003). Blood mercury levels in US children and women of childbearing age, 1999–2000. *The Journal of the American Medical Association, 289,* 1667–1674.

114. Trasande, L., Landrigan, P. J., & Schechter, C. (2005). Public health and economic consequences of methyl mercury toxicity to the developing brain. *Environmental Health Perspectives, 113,* 590–596.

115. Cohen, J. T., Bellinger, D. C., & Shaywitz, B. A. (2005). A quantitative analysis of prenatal methyl mercury exposure and cognitive development. *American Journal of Preventive Medicine, 29,* 353–365.

116. Schneider, H. J., et al. (2007). Hypopituitarism. *Lancet, 369,* 1461–1470.

117. Brown, K., et al. (2004). Abnormal cognitive function in treated congenital hypopituitarism. *Archives of Disease in Childhood, 89,* 827–830.

118. Frances, A., et al. (1994). *Diagnostic and Statistical Manual of Mental Disorders (DSM-IV)* (4th ed.). Washington DC: American Psychiatric Association.

119. Vasa, R. A., et al. (2006). Memory deficits in children with and at risk for anxiety disorders. *Depression and Anxiety, 24,* 85–94.

120. Domingues, R. B. et al. (2007). Involvement of the central nervous system in patients with dengue virus infection. Journal of the Neurological Sciences (E-pub ahead of print).

121. Solomon, T., et al. (2000). Neurological manifestations of dengue infection. *Lancet, 355,* 1053–1059.

122. Kalita, J., & Misra, U. K. (2006). EEG in degue virus infection with neurological manifestations: a clinical and CT/MRI correlation. *Clinical Neurophysiology, 117,* 2252–2256.

123. Seet, R. C., Quek, A. M., & Lim, E. C. (2007). Post-infectious fatigue syndrome in dengue infection. *Journal of Clinical Virology, 38,* 1–6.

124. Boswell, J. E., McErlean, M., & Verdile, V. P. (2002). Prevalence of traumatic brain injury in an ED population. *The American Journal of Emergency Medicine, 20,* 177–180.

125. Hawley, C. A., et al. (2004). Return to school after brain injury. *Archives of Disease in Childhood, 89,* 136–142.

126. Ewing-Cobbs, L., et al. (2006). Late intellectual and academic outcomes following traumatic brain injury sustained during early childhood. *Journal of Neurosurgery, 105,* 287–296.

127. Jonsson, C. A., Horneman, G., & Emanuelson, I. (2004). Neuropsychological progress during 14 years after severe traumatic brain injury in childhood and adolescence. *Brain Injury, 18,* 921–934.

128. McCarthy, J., et al. (2004). Sustained attention and visual processing speed in children and adolescents with bipolar disorder and other psychiatric disorders. *Psychological Reports, 95,* 39–47.

129. Centers for Disease Control. (2007). *Chagas Disease; Epidemiology and Risk Factors.* Washington, DC: Centers for Disease Control.

130. Mangone, C. A., et al. (1994). Cognitive impairment in human chronic Chagas' disease. *Arquivos de neuro-psiquiatria, 52*(2), 200–203.
131. Dye, C. (2006). Global epidemiology of tuberculosis. *Lancet, 367*, 938–940.
132. Trunz, B. B., & Dye, C. (2006). Effect of BCG vaccination on childhood tuberculous meningitis and miliary tuberculosis worldwide: a meta-analysis and assessment of cost-effectiveness. *Lancet, 367*, 1173–1180.
133. Schoeman, J. F., et al. (2002). Long-term follow up of childhood tuberculous meningitis. *Developmental Medicine and Child Neurology, 44*, 522–526.
134. Sarici, S. U., et al. (2004). Incidence, course, and prediction of hyperbilirubinemia in near-term and term newborns. *Pediatrics, 113*, 775–780.
135. Newman, T. B., et al. (2006). Outcomes among newborns with total serum bilirubin levels of 25 mg per deciliter or more. *The New England Journal of Medicine, 354*, 1889–1900.
136. Ip, S., et al. (2004). An evidence-based review of important issues concerning neonatal hyperbilirubinemia. *Pediatrics, 114*, e130–e153.
137. Wolf, M.-J., et al. (1999). Neurodevelopmental outcome at 1 year in Zimbabwean neonates with extreme hyperbilirubinemia. *European Journal of Pediatrics, 158*, 111–114.
138. Leimkugel, J., et al. (2007). Clonal waves of *Neisseria* colonisation and disease in the African meningitis belt: eight-year longitudinal study in northern Ghana. *PLoS Medicine, 4*, e101.
139. Pentland, L. M., Anderson, V. A., & Wrennall, J. A. (2000). The implications of childhood bacterial meningitis for language development. *Neuropsychology, Development, and Cognition. Section C Child Neuropsychology, 6*, 87–100.
140. van de Beek, D., et al. (2002). Cognitive impairment in adults with good recovery after bacterial meningitis. *The Journal of Infectious Diseases, 186*, 1047–1052.
141. Silvia, M. T., & Licht, D. J. (2005). Pediatric central nervous system infections and inflammatory white matter disease. *Pediatric Clinics of North America, 52*, 1107–1126.
142. Chen, K.-T., et al. (2007). Epidemiologic features of hand-foot-mouth disease and herpangina caused by enterovirus 71 in Taiwan, 1998–2005. *Pediatrics, 120*(2), e244–e252.
143. Chang, L.-Y., et al. (2007). Neurodevelopment and cognition in children with enterovirus 71 infection. *The New England Journal of Medicine, 356*, 1226–1234.
144. Robbins, J. M., et al. (2006). Reduction in newborns with discharge coding of *in utero* alcohol effects in the United States, 1993 to 2002. *Archives of Pediatrics & Adolescent Medicine, 160*(12), 1224–1231.
145. Forbess, J. M., et al. (2002). Neurodevelopmental outcome after congenital heart surgery: results from an institutional registry. *Circulation, 106*(Suppl I), I95–I102.
146. Greenaway, C. (2004). Dracunculiasis (Guinea worm disease). *Canadian Medical Association Journal, 170*, 495–500.
147. Ilegbodu, V. A., et al. (1986). Impact of guinea worm disease on children in Nigeria. *The American Journal of Tropical Medicine and Hygiene, 35*, 962–964.
148. Shah, P. M. (1991). Prevention of mental handicaps in children in primary health care. *Bulletin of the World Health Organization, 69*, 779–789.
149. Moreno, C. et al. (2007). National trends in the outpatient diagnosis and treatment of bipolar disorder in youth. *Archieves of General Psychiatry, 64*, 1032–1039.
150. McJunkin, J.E. et al. (2001) La Crosse encephalitis in children *New England Journal of Medicine, 344*, 801–807.

# Chapter 11

1. Walsh, B. T., et al. (2002). Placebo response in studies of major depression: Variable, substantial, and growing. *Journal of the American Medical Association, 287*, 1840–1847.
2. Steen, R. G. (2007). *The Evolving Brain: The Known and the Unknown* (p. 425). New York: Prometheus Books.

3.  Hochstenbach, J., et al. (1998). Cognitive decline following stroke: A comprehensive study of cognitive decline following stroke. *Journal of Clinical and Experimental Neuropsychology, 20*, 503–517.

4.  Meinzer, M., et al. (2005). Long-term stability of improved language functions in chronic aphasia after constraint-induced aphasia therapy. *Stroke, 36*, 1462–1466.

5.  Doesborgh, S. J. C., et al. (2004). Effects of semantic treatment on verbal communication and linguistic processing in aphasia after stroke: A randomized controlled trial. *Stroke, 35*, 141–146.

6.  Cicerone, K. D., et al. (2005). Evidence-based cognitive rehabilitation: Updated review of the literature from 1998 through 2002. *Archives of Physical Medicine and Rehabilitation, 86*, 1681–1692.

7.  Lincoln, N. B., et al. (1984). Effectiveness of speech therapy for aphasic stroke patients: A randomised controlled trial. *Lancet, 1*, 1197–1200.

8.  Pulvermuller, F., et al. (2001). Constraint-induced therapy of chronic aphasia after stroke. *Stroke, 32*, 1621–1626.

9.  Papathanasiou, I., et al. (2003). Plasticity of motor cortex excitability induced by rehabilitation therapy for writing. *Neurology, 61*, 977–980.

10. Nudo, R. J., et al. (1996). Neural substrates for the effects of rehabilitative training on motor recovery after iscemic infarct. *Science, 272*, 1791–1794.

11. Chilosi, A. M., et al. (2005). Atypical language lateralization and early linguistic development in children with focal brain lesions. *Developmental Medicine and Child Neurology, 47*, 725–730.

12. Sunderland, A., Stewart, F. M., & Sluman, S. M. (1996). Adaptation to cognitive deficit? An exploration of apparent dissociations between everyday memory and test performance late after stroke. *British Journal of Clinical Psychology, 35*, 463–476.

13. Peck, K. K., et al. (2004). Functional magnetic resonance imaging before and after aphasia therapy: Shifts in hemodynamic time to peak during an overt language task. *Stroke, 35*, 554–559.

14. Pulvermuller, F., et al. (2005). Therapy-related reorganization of language in both hemispheres of patients with chronic aphasia. *NeuroImage, 28*, 481–489.

15. Saur, D., et al. (2006). Dynamics of language reorganization after stroke. *Brain, 129*, 1371–1384.

16. Feeney, D. M. (2001). Changing foundations of speech therapy. *Stroke, 32*, 2097–2098.

17. Breitenstein, C., et al. (2004). D-Amphetamine boosts language learning independent of its cardiovascular and motor arousing effects. *Neuropsychopharm, 29*, 1704–1714.

18. Walker-Batson, D., et al. (2001). A double-blind, placebo-controlled study of the use of amphetamine in the treatment of aphasia. *Stroke, 32*, 2093–2098.

19. Breitenstein, C., et al. (2006). A shift of paradigm: From noradrenergic to dopaminergic modulation of learning? *Journal of Neurological Science, 248*, 42–47.

20. Martinsson, L., & Wahlberg, N. G. (2003). Safety of dexamphetamine in acute ischemic stroke: A randomized, double-blind, controlled dose-escalation trial. *Stroke, 34*, 475–481.

21. Mattay, V. S., et al. (2003). Catechol O-methyltransferase val-158-met genotype and individual variation in the brain response to amphetamine. *Proceedings of the National Academy of Sciences of the United States of America, 100*, 6186–6191.

22. Knutson, B., et al. (2004). Amphetamine modulates human incentive processing. *Neuron, 43*, 261–269.

23. Spencer, T. J., et al. (2006). Efficacy and safety of mixed amphetamine salts extended release (Adderall XR) in the management of attention-deficit/hyperactivity disorder in adolescent patients: A 4-week, randomized, double-blind, placebo-controlled, parallel-group study. *Clinical Therapy, 28*, 266–279.

24. Martinsson, L., Hardemark, H., & Eksborg, S. (2007). Amphetamines for improving recovery after stroke. *Cochrane Database of Systematic Reviews*, CD002090

25. Treig, T., et al. (2003). No benefit from D-amphetamine when added to physiotherapy after stroke: A randomized, placebo-controlled study. *Clinical Rehabilitation, 17*, 590–599.

26. Knecht, S., et al. (2001). D-Amphetamine does not improve outcome of somatosensory training. *Neurology, 57*, 2248–2252.

27. Martinsson, L., Eksborg, S., & Wahlberg, N. G. (2003). Intensive early physiotherapy combined with dexamphetamine treatment in severe stroke: A randomized, controlled pilot study. *Cerebrovascular Diseases, 16*, 338–345.

28. Platz, T., et al. (2005). Amphetamine fails to facilitate motor performance and to enhance motor recovery among stroke patients with mild arm paresis: Interim analysis and termination of a double blind, randomised, placebo-controlled trial. *Restorative Neurology and Neuroscience, 23*, 271–280.

29. Gladstone, D. J., et al. (2006). Physiotherapy coupled with dextroamphetamine for rehabilitation after hemiparetic stroke: A randomized, double-blind, placebo-controlled trial. *Stroke, 37*, 179–185.

30. Sonde, L., & Lokk, J. (2007). Effects of amphetamine and/or L-dopa and physiotherapy after stroke-a blinded randomized study. *Acta Neurologica Scandinavica, 115*, 55–59.

31. Paolucci, S., & De Angelis, D. (2006). New developments on drug treatment rehabilitation. *Clinical and Experimental Hypertension, 28*, 345–348.

32. Knecht, S., et al. (2004). Levodopa: Faster and better word learning in normal humans. *Annals of Neurology, 56*, 20–26.

33. Gayan, J., & Olson, R. K. (2003). Genetic and environment influences on individual differences in printed word recognition. *Journal of Experimental Child Psychology, 84*, 97–123.

34. Talcott, J. B., et al. (2000). Dynamic sensory sensitivity and children's word decoding skills. *Proceedings of the National Academy of Sciences of the United States of America, 97*, 2952–2957.

35. Tan, L. H., et al. (2005). Reading depends on writing, in Chinese. *Proceedings of the National Academy of Sciences of the United States of America, 102*, 8781–8785.

36. Stock, C. D., & Fisher, P. A. (2006). Language delays among foster children: Implications for policy and practice. *Child Welfare, 85*, 445–461.

37. Toppelberg, C. O., & Shapiro, T. (2000). Language disorders: A 10-year research update review. *Journal of the American Academy of Child and Adolescent Psychiatry, 39*, 143–152.

38. Miniscalco, C., et al. (2006). Neuropsychiatric and neurodevelopmental outcome of children at age 6 and 7 years who screened positive for language problems at 30 months. *Developmental Medicine and Child Neurology, 48*, 361–366.

39. Young, A. R., et al. (2002). Young adult academic outcomes in a longitudinal sample of early identified language impaired and control children. *Journal of Child Psychology Psychiatry, 43*, 635–645.

40. Singer-Harris, N., et al. (2001). Children with adequate academic achievement scores referred for evaluation of school difficulties: Information processing deficiencies. *Developmental Neuropsychology, 20*, 593–603.

41. Alston, E., & James-Roberts, I. S. (2005). Home environments of 10-month-old infants selected by the WILSTAAR screen for pre-language difficulties. *International Journal of Language and Communication Disorders, 40*, 123–136.

42. Horowitz, L., et al. (2005). Behavioral patterns of conflict resolution strategies in preschool boys with language impairment in comparison with boys with typical language development. *International Journal of Language and Communication Disorders, 40*, 431–454.

43. Ziegler, J. C., et al. (2005). Deficits in speech perception predict language learning impairment. *Proceedings of the National Academy of Sciences of the United States of America, 102*, 14110–14115.

44. Hatcher, P. J., et al. (2006). Efficacy of small group reading intervention for beginning readers with reading-delay: A randomised controlled trial. *Journal of Child Psychology and Psychiatry, 47*, 820–827.

45. Gillon, G. T. (2005). Facilitating phoneme awareness development in 3- and 4-year-old children with speech impairment. *Language, Speech, and Hearing Services in Schools, 36*, 308–324.

46. Churches, M., Skuy, M., & Das, J. P. (2002). Identification and remediation of reading difficulties based on successive processing deficits and delay in general reading. *Psychological Reports, 91*, 813–824.

47. Nancollis, A., Lawrie, B. A., & Dodd, B. (2005). Phonological awareness intervention and the acquisition of literacy skills in children from deprived social backgrounds. *Language, Speech, and Hearing Services in Schools, 36*, 325–335.

48. Strehlow, U., et al. (2006). Does successful training of temporal processing of sound and phoneme stimuli improve reading and spelling? *European Child and Adolescent Psychiatry, 15*, 19–29.

49. Merzenich, M. M., et al. (1996). Temporal processing deficits of language-learning impaired children ameliorated by training. *Science, 271*, 77–81.

50. Tallal, P., et al. (1996). Language comprehension in language-learning impaired children improved with acoustically modified speech. *Science, 271*, 81–84.

51. Cohen, W., et al. (2005). Effects of computer-based intervention through acoustically modified speech (Fast ForWord) in severe mixed receptive-expressive language impairment: Outcomes from a randomized controlled trial. *Journal of Speech, Language, and Hearing Research, 48*, 715–729.

52. Bishop, D. V., Adams, C. V., & Rosen, S. (2006). Resistance of grammatical impairment to computerized comprehension training in children with specific and non-specific language impairments. *International Journal of Language and Communication Disorders, 41*, 19–40.

53. Valentine, D., Hedrick, M. S., & Swanson, L. A. (2006). Effect of an auditory training program on reading, phoneme awareness, and language. *Perceptual and Motor Skills, 103*, 183–196.

54. Temple, E., et al. (2003). Neural deficits in children with dyslexia ameliorated by behavioral remediation: Evidence from functional MRI. *Proceedings of the National Academy of Sciences of the United States of America, 100*, 2860–2865.

55. Peterson, B. S., et al. (2002). A functional magnetic resonance imaging study of language processing and its cognitive correlates in prematurely born children. *Pediatrics, 110*, 1153–1162.

56. Hoeft, F., et al. (2006). Neural basis of dyslexia: A comparison between dyslexic and non-dyslexic children equated for reading ability. *Journal of Neuroscience, 26*, 10700–10708.

57. Vigneau, M., et al. (2006). Meta-analyzing left hemisphere language areas: Phonology, semantics, and sentence processing. *Neuroimage, 30*, 1414–1432.

58. Booth, J. R., et al. (2007). Children with reading disorder show modality independent brain abnormalities during semantic tasks. *Neuropsychologia, 45*, 775–783.

59. Goorhuis-Brouwer, S., & Kniff, W. A. (2002). Efficacy of speech therapy in children with language disorders: Specific language impairment compared with language impairment in comorbidity with cognitive delay. *International Journal of Pediatric Otorhinolaryngology, 63*, 129–136.

60. Simos, P. G., et al. (2007). Intensive instruction affects brain magnetic activity associated with oral word reading in children with persistent reading disabilities. *Journal of Learning Disabilities, 40*, 37–48.

61. Vance, M., Stackhouse, J., & Wells, B. (2005). Speech-production skills in children aged 3–7 years. *International Journal of Language and Communication Disorders, 40*, 29–48.

62. Bernhardt, B., & Major, E. (2005). Speech, language and literacy skills 3 years later: A follow-up study of early phonological and metaphonological intervention. *International Journal of Language and Communication Disorders, 40*, 1–27.

63. Russo, N. M., et al. (2005). Auditory training improves neural timing in the human brainstem. *Behavioural Brain Research, 156*, 95–103.

64. Alexander, A. W., & Slinger, A. M. (2004). Current status of treatments for dyslexia: Critical review. *Journal of Child Neurology, 19*, 744–758.

65. Strakowski, S. M., & Sax, K. W. (1998). Progressive behavioral response to repeated D-amphetamine challenge: Further evidence for sensitization in humans. *Biological Psychiatry, 44*, 1171–1177.

66. Aldenkamp, A. P., et al. (2006). Optimizing therapy of seizures in children and adolescents with ADHD. *Neurology, 67*(Suppl 4), S49–S51.

67. Steen, R. G., & Mirro, J. (2000). *Childhood Cancer: A Handbook from St. Jude Children's Research Hospital*, Cambridge, MA: Perseus Publishing, 606 pp.

68. Steen, R. G., & Campbell, F. A. (2008). The cognitive impact of systemic illness in childhood and adolescence. In L. Weiss & A. Prifitera (Eds.), *WISC-IV: Clinical Use and Interpretation.* New York: Academic Press.

69. Thompson, S. J., et al. (2001). Immediate neurocognitive effects of methylphenidate on learning-impaired survivors of childhood cancer. *Journal of Clinical Oncology, 19,* 1802–1808.

70. Mulhern, R. K., et al. (2004). Short-term efficacy of methylphenidate: A randomized, double-blind, placebo-controlled trial among survivors of childhood cancer. *Journal of Clinical Oncology, 22,* 4795–4803.

71. Williams, S. E., et al. (1998). Recovery in pediatric brain injury: Is psychostimulant medication beneficial? *The Journal of Head Trauma Rehabilitation, 13,* 73–81.

72. Tillery, K. L., Katz, J., & Keller, W. D. (2000). Effects of methylphenidate (Ritalin) on auditory performance in children with attention and auditory processing disorders. *Journal of Speech, Language, and Hearing Research, 43,* 893–901.

73. Mahalick, D. M., et al. (1998). Psychopharmacologic treatment of acquired attention disorders in children with brain injury. *Pediatric Neurosurgery, 29,* 121–126.

74. Greenhill, L. L., Findling, R. L., & Swanson, J. M. (2002). A double-blind, placebo-controlled study of modified-release methylphenidate in children with attention-deficit/hyperactivity disorder. *Pediatrics, 109,* E39–E46.

75. Silva, R., et al. (2004). Open-label study of dexmethylphenidate hydrochloride in children and adolescents with attention deficit hyperactivity disorder. *Journal of Child and Adolescent Psychopharmacology, 14,* 555–563.

76. Wigal, S., et al. (2004). A double-blind, placebo-controlled trial of dexmethylphenidate hydrochloride and d, l-threo-methylphenidate hydrochloride in children with attention-deficit/hyperactivity disorder. *Journal of the American Academy of Child and Adolescent Psychiatry, 43,* 1406–1414.

77. Wigal, S. B., Gupta, S., & Greenhill, L. (2007). Pharmacokinetics of methylphenidate in preschoolers with attention-deficit/hyperactivity disorder. *Journal of Child and Adolescent Psychopharmacology, 17,* 153–164.

78. Arnold, L. E., et al. (2004). A double-blind, placebo-controlled withdrawal trial of dexmethylphenidate hydrochloride in children with attention deficit hyperactivity disorder. *Journal of Child and Adolescent Psychopharmacology, 14,* 542–554.

79. Bedard, A. C., et al. (2004). Methylphenidate improves visual-spatial memory in children with attention-deficit/hyperactivity disorder. *Journal of the American Academy of Child and Adolescent Psychiatry, 43,* 260–268.

80. Kopecky, H., et al. (2005). Performance and private speech of children with attention-deficit/hyperactivity disorder while taking the Tower of Hanoi test: Effects of depth of search, diagnostic subtype, and methylphenidate. *Journal of Abnormal Child Psychology, 33,* 625–638.

81. Evans, S. W., et al. (2001). Dose-response effects of methylphenidate on ecologically valid measures of academic performance and classroom behavior in adolescents with ADHD. *Experimental and Clinical Psychopharmacology, 9,* 163–175.

82. Bedard, A. C., Ickowicz, A., & Tannock, R. (2002). Methylphenidate improves Stroop naming speed, but not response interference, in children with attention deficit hyperactivity disorder. *Journal of Child and Adolescent Psychopharmacology, 12,* 301–309.

83. Francis, S., Fine, J., & Tannock, R. (2001). Methylphenidate selectively improves story retelling in children with attention deficit hyperactivity disorder. *Journal of Child and Adolescent Psychopharmacology, 11,* 217–228.

84. Swanson, J. M., et al. (2006). Modafinil film-coated tablets in children and adolescents with attention-deficit/hyperactivity disorder: Results of a randomized, double-blind, placebo-controlled, fixed-dose study followed by abrupt discontinuation. *Journal of Clinical Psychiatry, 67,* 137–147.

85. Greenhill, L. L., et al. (2006). A randomized, double-blind, placebo-controlled study of modafinil film-coated tablets in children and adolescents with attention-deficit/hyperactivity disorder. *Journal of the American Academy of Child and Adolescent Psychiatry, 45,* 503–511.

86. Boellner, S. W., Earl, C. Q., & Arora, S. (2006). Modafinil in children and adolescents with attention-deficit/hyperactivity disorder: A preliminary 8-week, open-label study. *Current Medical Research and Opinion, 22,* 2457–2465.
87. Schroeder, L., et al. (2006). The economic costs of congenital bilateral permanent childhood hearing impairment. *Pediatrics, 117,* 1101–1112.
88. Wilson, S. J., & Lipsey, M. W. (2003). The effects of school-based intervention programs on aggressive behavior: A meta-analysis. *Journal of Consulting and Clinical Psychology, 71,* 136–149.
89. Census Bureau, U.S. (2005). Median earnings in the past 12 months (in 2005 inflation-adjusted dollars) by sex by educational attainment for the population 25 years and over. Washington, DC

# Chapter 12

1. Steen, R. G., & Campbell, F. A. (2008). The cognitive impact of systemic illness in childhood and adolescence. In L. Weiss & A. Prifitera (Eds.), *WISC-IV: Clinical Use and Interpretation* (pp. 365–407). New York: Academic Press.
2. New York Times (2007). Editorial, Counting the poor, New York.
3. Office of Head Start. (2007). *Head Start Program Fact Sheet.* Washington, DC: Administration for Children & Families.
4. Puma, M., et al. (2005). Head Start Impact Study: First Year Findings. In Administration for Children & Families (ed.). U.S. Department of Health and Human Services: Washington, DC. p. 333.
5. Wasik, B. A., Bond, M. A., & Hindman, A. (2006). The effects of a language and literacy intervention on Head Start children and teachers. *Journal of Educational Psychology, 98,* 63–74.
6. Malofeeva, E., et al. (2004). Construction and evaluation of a number sense test with Head Start children. *Journal of Educational Psychology, 96,* 648–659.
7. Mantzicopoulos, P. (2003). Flunking kindergarten after Head Start: An inquiry into the contribution of contextual and individual variables. *Journal of Educational Psychology, 95,* 268–278.
8. Gunn, B., et al. (2006). Promoting school success: Developing social skills and early literacy in Head Start classrooms. *NHSA Dialog, 9,* 1–11.
9. Lipsey, M. W., & Wilson, D. B. (1993). The efficacy of psychological, educational, and behavioral treatment: Confirmation from meta-analysis. *American Psychologist, 48,* 1181–1209.
10. Love, J. M., et al. (2005). The effectiveness of Early Head Start for 3-year-old children and their parents: Lessons for policy and programs. *Developmental Psychology, 41,* 885–901.
11. St. Pierre, R. G., Ricciuti, A. E., & Rimdzius, T. A. (2005). Effects of a family literacy program on low-literate children and their parents: Findings from an evaluation of the Even Start Family Literacy Program. *Developmental Psychology, 41,* 953–970.
12. Olds, D. L., Sadler, L., & Kitzman, H. (2007). Programs for parents of infants and toddlers: Recent evidence from randomized trials. *Journal of Child Psychology and Psychiatry, 48,* 355–391.
13. Martin, S. L., Ramey, C. T., & Ramey, S. (1990). The prevention of intellectual impairment in children of impoverished families: Findings of a randomized trial of educational day care. *American Journal of Public Health, 80,* 844–847.
14. Campbell, F. A. (2009). Early childhood interventions: The Abecedarian Project. (in prep.).
15. Campbell, F. A., et al. (2001). The development of cognitive and academic abilities: Growth curves from an early childhood educational experiment. *Developmental Psychology, 37,* 231–242.
16. Sattler, J. M. (1992). Assessment of ethnic minority children. In *Assessment of Children* (pp. 563–596). Jerome M. Sattler, Publisher, Inc.: San Diego.

17. Campbell, F. A., & Ramey, C. T. (1994). Effects of early intervention on intellectual and academic achievement: A follow-up study of children from low-income families. *Child Development, 65*, 684–698.
18. Burchinal, M. R., et al. (1997). Early intervention and mediating processes in cognitive performance of children of low-income African-American families. *Child Development, 68*, 935–954.
19. Ramey, C. T., et al. (2000). Persistent effects of early childhood education on high-risk children and their mothers. *Applied Developmental Science, 4*, 2–14.
20. McLaughlin, A. E., et al. (2007). Depressive symptoms in young adults: The influences of the early home environment and early educational child care. *Child Development, 78*, 746–756.
21. Campbell, F. A., et al. (2002). Early childhood education: Young adult outcomes from the Abecedarian Project. *Applied Developmental Science, 6*, 42–57.
22. Behrman, R. E. (ed). (1995). *Long-term outcomes of early childhood programs. The future of children, vol. 5(3)*. Princeton, NJ: Princeton University – The Brookings Institution.
23. Waldfogel, J. (1999). *Early childhood interventions and outcomes*. London: Center for Analysis of Social Exclusion, London School of Economics.
24. Feder, L. & Clay J.A. (2002) *Investing in the future of our children: Building a model community prevention program*. Memphis. http://suds.memphis.edu/reports/Ecca_best_practices.pdf
25. Donovan, M. S. & Cross C. T. (2002). Minority Students in Special and Gifted Education. Division of Behavioral and Social Sciences and Education, Committee on Minority Representation in Special Education (ed.) Washington, DC: National Academy Press.
26. Wise, S., et al. (2005) The efficacy of early childhood interventions. Australian Government Department of Family and Community Services (ed.) Melbourne, Australia: Melbourne Institute of Applied Economic and Social Research.
27. Anderson, M. L. (2005) Uncovering gender differences in the effects of early intervention: A reevaluation of the Abecedarian, Perry Preschool, and Early Training Projects. Social Science Research Network.
28. Hamm, K. & Ewen D. (2006) From the beginning: Early Head Start children, families, staff, and programs in 2004. In Brief C. P. (ed.) Center for Law and Social Policy.
29. Johnson, D. L. (2006). Parent–Child Development Center follow-up project: Child behavior problem results. *The Journal of Primary Prevention, 27*, 391.
30. Garber, H.L., The Milwaukee Project: Preventing Mental Retardation in Children at Risk. (1988). *Washington*. DC: American Association on Mental Retardation.
31. Weikart, D. P., Bond J. T., & McNeil J. T. (1978). The Ypsilanti Perry Preschool Project: Preschool years and longitudinal results through fourth grade. In *Monographs of the High Scope Educational Research Foundation*. High Scope Press: Ypsilant, MI.
32. Reynolds, A. J., et al. (2001). Long-term effects of an early childhood intervention on educational achievement and juvenile arrest: a 15-year follow-up of low-income children in public schools. *JAMA, 285*, 2339–2346.
33. Palfrey, J. S., et al. (2005). The Brookline Early Education Project: A 25-year follow-up study of a family-centered early health and development intervention. *Pediatrics, 116*, 144–152.
34. Zigler, E., & Styfco, S. J. (2001). Extended childhood intervention prepares children for school and beyond. *JAMA, 285*, 2378–2380.

# Chapter 13

1. Barker, D. J. P., et al. (2005). Infant growth and income 50 years later. *Archives of Disease in Childhood, 90*, 272–273.
2. Steen, R. G., & Campbell, F. A. (2008). The cognitive impact of systemic illness in childhood and adolescence. In L. Weiss & A. Prifitera (Eds.), *WISC-IV: Clinical Use and Interpretation* (pp. 365–407). New York: Academic Press.

3. Campbell, F. A., et al. (2002). Early childhood education: Young adult outcomes from the Abecedarian Project. *Applied Developmental Science, 6*, 42–57.

4. Reynolds, A. J., et al. (2001). Long-term effects of an early childhood intervention on educational achievement and juvenile arrest: A 15-year follow-up of low-income children in public schools. *Journal of the American Medical Association, 285*, 2339–2346.

5. Palfrey, J. S., et al. (2005). The Brookline Early Education Project: A 25-year follow-up study of a family-centered early health and development intervention. *Pediatrics, 116*, 144–152.

6. Belsky, J., et al. (2007). Are there long-term effects of early child care? *Child Development, 78*, 681–701.

7. Lansford, J. E., et al. (2002). A 12-year prospective study of the long-term effects of early child physical maltreatment on psychological, behavioral, and academic problems in adolescence. *Archives of Pediatrics and Adolescent Medicine, 156*, 824–830.

8. Steen, R. G. (1996). *DNA and Destiny: Nature and Nurture in Human Behavior* (p. 295). New York: Plenum Press.

9. Kramer, R. A., Allen, L., & Gergen, P. J. (1995). Health and social characteristics and children's cognitive functioning: Results from a national cohort. *American Journal of Public Health and the Nation's Health, 85*, 312–318.

10. Currie, J. (2005). Health disparities and gaps in school readiness. *Future Child, 15*, 117–138.

11. Hillis, S. D., et al. (2004). The association between adverse childhood experiences and adolescent pregnancy, long-term psychosocial consequences, and fetal death. *Pediatrics, 113*, 320–327.

12. Kahn, R. S., Wilson, K., & Wise, P. H. (2005). Intergenerational health disparities: Socioeconomic status, women's health conditions, and child behavior problems. *Public Health Reports, 120*, 399–408.

13. Hobcraft, J., & Kiernan, K. (2001). Childhood poverty, early motherhood and adult social exclusion. *The British Journal of Sociology, 52*, 495–517.

14. Belsky, J., et al. (2007). Socioeconomic risk, parenting during the preschool years and child health age 6 years. *European Journal of Public Health, 12*, 508–513.

15. Davis-Kean, P. E. (2005). The influence of parent education and family income on child achievement: The indirect role of parental expectations and the home environment. *Journal of Family Psychology, 19*, 294–304.

16. Finkelstein, D. M., et al. (2007). Socioeconomic differences in adolescent stress: The role of psychological resources. *The Journal of Adolescent Health , 40*, 127–134.

17. Kolata, G. (2007). A surprising secret to a long life: stay in school. *New York Times,* New York.

18. Leonhardt, D. (2007). What $1.2 trillion can buy. *New York Times,* New York.

19. Kennedy, E. M. (2004). Keeping faith with our children. The American Prospect (November): p. A2–A3

20. Msall, M. E. (2004). Supporting vulnerable preschool children: Connecting the dots before kindergarten. *Pediatrics, 114*, 1086.

21. Ludwig, J., & Phillips, D. A. (2008). Long-term effects of Head Start on low-income children. *Annals of the New York Academy of Sciences, 1136*, 1–12.

22. Dillon, S. (2006). Schools slow in closing gaps between races. *New York Times,* New York.

23. Dillon, S. (2008). State's data obscure how few finish high school. *New York Times,* New York.

24. FaithfulAmerica website. 2007.

25. Knudsen, E. I., et al. (2006). Economic, neurobiological, and behavioral perspectives on building America's future workforce. *Proceedings of the National Academy of Sciences of the United States of America, 103*, 10155–10162.

26. Blank, H. (2004). Head Start under assault. *The American Prospect* 2004(November): A10–A11.

27. National Education Association website. (2007) *Teacher salaries.* from http://www.nea.org/neatodayextra/salaries.html.

28. Hungerford, T. L., & Wassmer, R. W. (2004). *K-12 education in the US economy: Its impact on economic development, earnings, and housing values. NEA Research Working Paper.* Washington, DC: NEA.

29. Dillon, S. (2007). Long reviled, merit pay gains among teachers. *New York Times,* New York.
30. Barnett, W. S. (1998). Long-term cognitive and academic effects of early childhood education on children in poverty. *Preventive Medicine, 27,* 204–207.
31. Heckman, J. J. (2006). Skill formation and the economics of investing in disadvantaged children. *Science, 312,* 1900–1902.
32. Kirp, D. L. (2004). You're doing fine, Oklahoma! *The American Prospect.* (November): p. A5–A8
33. Leonhardt, D. (2007). Bridging gaps early on in Oklahoma. *New York Times,* New York.
34. Campbell, F. A., et al. (2001). The development of cognitive and academic abilities: Growth curves from an early childhood educational experiment. *Developmental Psychology, 37,* 231–242.
35. Shonkoff, J. P. (2005). The non-nuclear option. *The American Prospect, 16*(5), A18–A20.

# Index